Hi

Flora

Because it is one of the great unspoiled areas of Europe, the Highlands and Islands are able to put on many of nature's most colourful and lively displays. Nowhere is this more true than in the case of flora. Fragile alpines, sturdy oak and, in places, even plants whose natural home is in warmer climes—all contribute to the mosaic of the area's vegetable life.

Derek Ratcliffe, from years rich in experience and research, knows the Highlands and Islands well. He brings all of this to bear on a botanical journey round the area in a way which will inspire the knowledgeable and enthuse the interested.

Highland Life Series

Highland Flora
by Derek Ratcliffe

Published by the
Highlands and Islands Development Board
Bridge House, Bank Street, Inverness

First published 1977

Also in the Highland Life Series
Highland Birds
Highland Animals
Highland Landforms

Designed by
Mackay Design Associates Ltd

Printed by Morrison & Gibb Ltd
Edinburgh and London

ISBN 0 902347 56 X

All the illustrations in this book apart from those listed below were provided by the author: Derek Ratcliffe.

The use of material provided by the following is gratefully acknowledged: R M Adam: Figs 2, 18, 22, 24, 26, 32, 65, 68. H J B Birks: Figs 27, 33, 43, 50, 59, Front Cover 1, Inside Cover. D Gowans: Figs 8, 47, 52, Front 3–5 and Back Cover. D Hayes: Fig 38. M C F Proctor: Figs 10, 11, 19, 20, 21, 23, 25, 34, 42, 53, 69, 70, 71. J G Roger: Figs 28, 35, 51.

Also in the Highland Life series:

Highland Birds
by Desmond Nethersole-Thompson
First Published 1971: second edition published 1974

For over 30 years the author has lived and worked in the Highlands. In this intimate personal sketch Desmond Nethersole-Thompson demonstrates his love for Highland Birds.

Highland Animals
by David Stephen
First Published 1974

David Stephen has been studying mammals for more than 40 years, and has done a considerable amount of research on many species. Author of the best seller 'Watching Wildlife', he reveals to the reader of Highland Animals his personal excitement and interest in his subject.

Highland Landforms
by Robert J Price
First Published 1976

Robert J Price has devoted 20 years to the forces that shape 'planet earth'. In Highland Landforms, a personal examination of the territory lying between Shetland and the Firth of Clyde, he communicates to his readers some of the excitement produced by what is the most freshly minted landscape in north-west Europe.

Copies of all titles are available from bookshops or direct from Publications, Highlands and Islands Development Board, Bridge House, 27 Bank Street, Inverness IV1 1QR. Write for details of prices.

Dr. D A Ratcliffe
Biographical Details

Born: London, but grew up in Carlisle, Cumbria

Educated at University of Sheffield, B.Sc., D.Sc.
University of Wales Ph.D.

Research biologist and nature conservationist specialising in plant ecology and ornithology, with a particular interest in wildlife of mountains and moorlands. Joint author of a monograph on vegetation of the Scottish Highlands.

Current position: Chief Scientist of the Nature Conservancy Council.

Front Cover
1. Mountain and loch: Sgurr nan Gillean, Cuillins from near Sligachan, Skye.
2. Scottish primrose: one of the few plants confined to Scotland.
3. Yellow flag (iris): this marsh plant grows in profusion along many western and northern shores.
4. Common cotton grass: the attractive fruiting heads of this abundant wetland plant.
5. Deer sedge: an extremely abundant plant of the northern moorlands.
Inside cover – Hay-scented fern, a local, warmth-loving fern common in some rocky woods and slopes along the Atlantic seaboard.
Back cover – The Cairngorms from the Rothiemurchus pinewoods.

Contents

List of Illustrations

Preface

This is the fourth title in our Highland Life series and, but for one sector, completes what may be called the naturalist's view of the Highlands and Islands. The gap in the library relates to the water resources of the area and that we hope to cover in time.

"Highland Flora" offered the author, Derek Ratcliffe, something of a challenge. That he has met it in such a colourful and interesting way, I am sure, will benefit all those who read this book. It is no easy task to assess all of the Highlands and Islands from one particular viewpoint but Dr Ratcliffe has gone about it with a relish which will make Highland Flora a real addition to the Highland Life library.

To him and to all those who helped him, the Board owes a debt of thanks.

R. Fasken

Board Member

Foreword

The Highlands and Islands of Scotland contain over 1,500 different kinds of wild flowering plants and ferns, and a still greater number of lowlier forms. Plants grow almost everywhere within this large and varied region except on completely artificial surfaces and unweathered rocks. Much of the scenic character and aesthetic appeal of any region depends on the prevailing mantle of vegetation, both natural and man-made, which gives colour and texture to the landscape, and enhances its variety of form. Most people are more aware of plant life in its essentially individual aspects—the beauty of the trees, shrubs and wild flowers in Nature's garden. I have tried to portray both the larger and the smaller scales of botanical diversity, and to relate one to the other. Although the Highlands and Islands are the largest area of undeveloped land in Britain, and their flora consists mainly of truly native plants, the way in which these plants grow together to form vegetation has been profoundly influenced by human activity. Plant distribution may at first sight seem to show a confusing or even chaotic picture, but it is far from random, and there is a certain pattern and orderliness which can be observed and understood. It is my aim to sketch out and explain this order in outline, whilst attempting also to convey something of the very direct appeal which the Highland flora makes to the senses.

Desmond Nethersole-Thompson first suggested such a book to me, and I owe a great deal to his encouragement and enthusiasm for the project. The Highlands and Islands Development Board invited me to write Highland Flora as the fourth of their Highland Life series and generously gave me a free hand in its compilation. I am grateful to James Grassie and Gordon Lyall of HIBD, and to Bill Mackay and his colleagues in Mackay Design Associates for their help and guidance at various stages, but especially in the final selection of the illustrations. And I acknowledge with much appreciation the contributions of the other photographers: Michael Proctor, John Birks, David Gowans, Grant Roger, and David Hayes. Messrs. D C Thomson and Co. Ltd., of Dundee kindly made available photographs from the collection of the late Robert Adam, and I thank Mr D M Spence, their Photographic Librarian.

As regards content I am indebted to Harold Fletcher for allowing me to draw from his paper on historical aspects of Highland botany. I have made variable and arbitrary reference to the work of living botanists, and hope that the many whose names are omitted will not find this too capricious a treatment. It is, in fact, too short a book to do justice to the dedicated labours of the great many botanists, past and present, whose combined efforts have built up our knowledge of Highland plants. I can but pay tribute to them collectively, and regret that a roll of honour is not possible here.

I thank those good friends who have given their knowledge and companionship during many an enjoyable day in the field—Grant Roger, Donald McVean, John and Hilary Birks, and John Mitchell. Several of my colleagues in the Nature Conservancy Council have kindly read and commented upon the text, but the views expressed therein are my own and should not be taken to reflect official attitudes of the Council.

Finally, it is with great pleasure that I thank Jeannette Chan-Mo, who gave valuable help in preparation of the manuscript.

Derek Ratcliffe

Botanising in the Highlands

The plant life of the Highlands and Islands of Scotland is rich and fascinating in variety, ranging as it does from the lowly seaweeds and toadstools to the stately and ancient trees of the forests. Perhaps it is the beauty of the flowering plants which has the greatest appeal, and even though there are not the dazzling floral displays of the Alps, the region has much to delight the traveller by way of botanical treats. The mantle of both natural and man-made vegetation gives the Highlands much of their special character and charm. Some common wild plants make a striking contribution to the colours of the scene, first among them being the ling heather which in August covers whole hillsides and moors with a blaze of purple. Earlier in the year luminous yellow thickets of gorse and broom are another great feature along many roadsides and lower slopes (*Fig. 2*).

Each month has something new to offer in the pattern of colour and tone. In the drabness of winter the unchanging dark-green beauty of the Scots pine is the more clearly revealed, and there is a great richness of colour in the tawny swathes of dead bracken when lit by the low rays of an evening sun. There is special grace in the spring-time birches as they burst into delicate pale green, and a contrasting splendour in the autumnal russet shades of oak.

My aim in this book is to convey something of these visual qualities of the wild plants of the Highlands and Islands through the pictures, which speak for themselves, and to give some account of their distribution and how they came to be where they are. It seeks to add to the direct aesthetic rewards of seeing these lovely ornaments of nature in their natural setting by giving insight into the influences which shape their existence—the climate, rocks and soils, landforms and human activities. It is not a comprehensive guide to the identification of all the plants which may be encountered in the Highlands and Islands, though the photographs will naturally help in the recognition of those species which are illustrated. For plant identification an excellent range of books is available, from the simple to the technical, and a short list of some useful titles is given in the selected bibliography.

Nor is this book intended to give precise directions for locating the less common plants. Part of the fun and sense of achievement in botanising comes from finding the rarer species for oneself. For myself, there is more enjoyment in being a plant-hunter than a list-ticker—a preference for getting off the beaten track and exploring rather than heading straight for the famous places of pilgrimage. In consequence there are several Highland flowering plants which I have not yet seen, but I have had the pleasure of seeing quite a number of others for the first time in places where they were previously unknown.

The common and widespread plants can be seen with little or no effort. Of late the chair-lifts and mountain roads have brought many of the alpines within easy access of those who would otherwise have difficulty in reaching the high ground. I shall try to give a description of the more distinctive botanical habitats within each district of the Highlands and Islands. There is a good deal of overlap between these districts, which it would be unduly repetitive to attempt to portray. A brief generalised account of vegetation has therefore been provided to help the reader to interpret how widely or otherwise the description of a particular area, such as Rannoch Moor, can be applied to similar kinds of terrain elsewhere. No attempt has been made to pin-point the localities for extreme rarities, as this would not be in the best interests of these plants, some of which have suffered from far too much attention already. Indeed, to those who have the good fortune to chance upon new discoveries of rare plants, I would urge the need for the utmost discretion in telling of these finds.

The greatest threat to the survival of this fine flora comes, sadly, from the activities of human beings themselves. Many beautiful plants have been entirely lost or woefully reduced by the incidental effects of the ways in which Man has used the land, through the destruction of original forests, cultivation, heavy grazing and the use of fire. But once species became rare, either through climatic restriction or human agency, they developed a special fascination for plant collectors.

Virtually all the early botanists whose explorations put the Highland flora on the map were avid collectors. Many of them behaved without a thought to the effects that their removal of rare plants were having on the continued existence of these species. Certain localities became famous, and generation after generation of plant-seekers converged upon them, all coming away with the cherished trophies in their collecting boxes. Some of the most sought after kinds were in sufficient quantity to withstand this human grazing, but others became greatly depleted. Visit any of the numerous major botanical institutions in this country and ask to see the herbarium, with its carefully pressed and mounted collection of dried plants. You will be sure to find the same rarities, from the same places, over and over again with depressing regularity: alpine forget-me-not, snow gentian and drooping saxifrage from Ben Lawers, yellow oxytropis and oblong woodsia from Corrie Fee, and so on. It is small wonder that for some alpines, there is probably more material stuck on sheets of paper as discoloured, brittle relics, than now survives in the wild state.

The time when there was an excuse for this magpie collecting of rare plants for the private herbarium has long passed. This kind of collection has waned in popularity anyway but not so the taking of living plants to grow. In some respects the gardeners, both amateur and professional, have done still more damage, for so often they uproot their plants.

It is indeed a great temptation to take back specimens of beautiful alpines and other plants to try them in the garden, but please try hard to resist. Many of the native mountain plants need special conditions of climate and soil which they will not find in the suburban rock garden, and will fade away when transplanted there. Many nurserymen sell acclimatised stock of most of our more attractive species, and this will have a much better chance of success than wild material uprooted and carted around in the car boot in polythene bags. For those who cannot resist the urge to take back holiday mementos of this kind, try the seed or cuttings and not whole plants torn up by the roots.

This is why there has to be such secrecy about the rarest plants, until such time as everyone truly believes that a wild plant should be left where it belongs, to be admired and respected by all who will pass that way, and for long after those now living have all vanished from the scene.

These living things are part of what has been called 'The Heritage of Wild Nature,' held in trusteeship by one generation for those to follow. There is now in any case a law forbidding the picking and uprooting of wild plants without the consent of the owner or occupier of the land on which they grow. A select 21 species of either exceptional rarity or beauty, or both, are protected by special penalties. These specially protected plants which grow in the Highlands are the Killarney fern, the alpine and oblong woodsias, Menzies' heath, Diapensia, drooping saxifrage, tufted saxifrage, snow gentian and alpine sow-thistle.

There is, however, a substitute form of collecting which is both challenging and rewarding, and that is with the camera. In these days of 35 mm single lens reflex cameras and colour film, there is great scope for plant photography, especially in such splendid settings as the Highlands and Islands can offer. The colour transparency or print can fix the fleeting

National Nature Reserves in the HIDB Area

	Reserve	*Region*	*Acres*
1	'Allt nan Carnan	Highland Region	18
2	Beinn Eighe	Highland Region	11,757
3	Ben Lui (part)	Strathclyde Region	1,210
4	Cairngorms (part)	Highland Region	41,382
5	Corrieshalloch	Highland Region	13
6	Craigellachie	Highland Region	642
7	Glasdrum Wood	Strathclyde Region	42
8	Glen Diomhan	Strathclyde Region	24
9	Glen Roy	Highland Region	2,887
10	Gualin	Highland Region	6,232
11	Haaf Gruney	Shetland Islands Area	44
12	Hermaness	Shetland Islands Area	2,383
13	Inchnadamph	Highland Region	3,200
14	Invernaver	Highland Region	1,363
15	Inverpolly	Highland Region	26,827
16	Keen of Hamar	Shetland Islands Area	75
17	Loch Druidibeg	Western Isles Islands Area	4,145
18	Monach Isles	Western Isles Islands Area	1,425
19	Mound Alderwoods	Highland Region	659
20	North Rona and Sula Sgeir	Western Isles Islands Area	320
21	Noss	Shetland Islands Area	774
22	Rassal Ashwood	Highland Region	209
23	Rhum	Highland Region	26,400
24	St. Kilda	Western Isles Islands Area	2,107
25	Strathy Bog	Highland Region	120

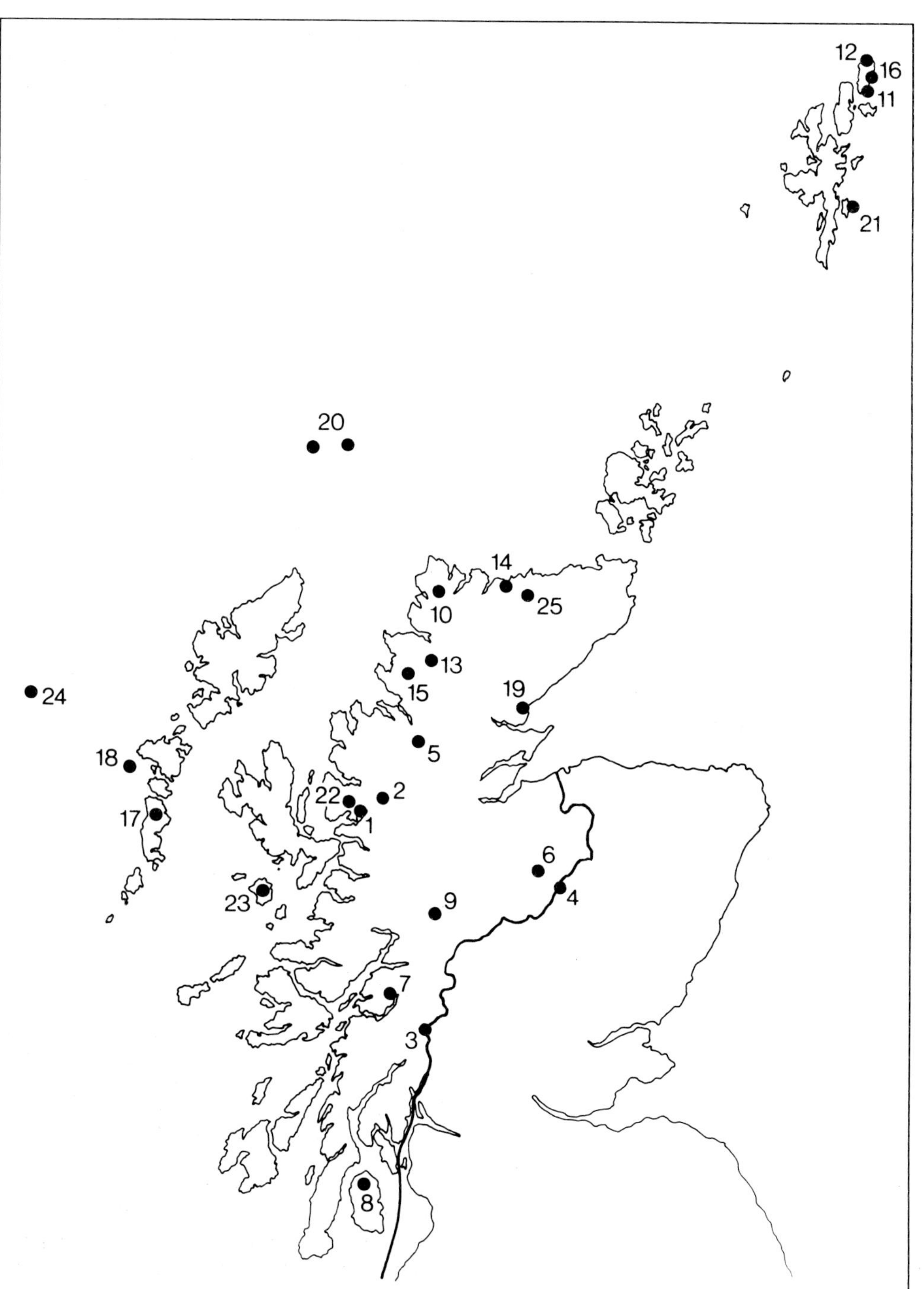

moments of beauty and fill out the memories with reality during moments of personal reminiscence or when trying to convey one's experiences to others. Yet even this is not an entirely harmless pursuit when people are thoughtless or careless. Direct damage can be done by trampling or over-zealous 'gardening' when actually taking the picture, and sometimes the owners or custodians of land with well-known rare plants have become so tired of the constant disturbance and even damage caused by a procession of visitors that they have closed access and, in one extreme case, threatened to eradicate the plant itself.

The photography itself has to be learned by experience, for there are certain difficulties. Wind is the great enemy, and some frail flowers continue to nod and sway when there is scarcely a perceptible movement of the air. And on those rare occasions when everything becomes absolutely still, out come the midges in swarms to try the patience of the most determined picture taker. Some photographers use flash to overcome the wind problem, but there is often a degree of harshness and shadow unless the sophisticated arrangement of a twin flash unit is used. A tripod is essential to achieve the best results. There are various texts on the subject to help the beginner and that by Michael Proctor (see bibliography) is particularly recommended.

The wild plants of the Highlands are everywhere, but some people look for guidance on how to become acquainted. Nature trails are both numerous and popular, but can seldom deal with more than a small proportion of the flora along their route. The National Trust for Scotland owns some properties with high botanical interest and has set up information centres at a few of these, such as Ben Lawers and Glen Torridon. The Nature Conservancy Council has established 25 National Nature Reserves within the Highlands, and these contain examples of most of the major vegetation types, each with its own characteristic assemblage of wild plants. Some of these reserves were chosen for the richness of their flora, including the presence of rare and attractive plants. Access to the reserves ranges from wholly unrestricted to entry by written permit only.

Restrictions are necessary partly because many reserves are not owned by the Council, and their owners required limitations on access, and partly because the needs of conservation of the flora and fauna may require some control of human activity on the site. Many reserves are under the charge of resident wardens who are the focus for information on these places, but whose many-sided duties often prevent them from being generally available to act as personal guides. The National Nature Reserves act partly as outdoor museums within which the public can learn about and enjoy wildlife, but they also act as refuges and reservoirs for many sensitive and sometimes threatened plants and animals which require quite strict protection if they are to survive. Some, too, are important research areas where scientific studies and experiments can be conducted with a minimum of disturbance. The reserves thus cannot be places where the public can desport themselves at will, as in the country parks, but need more regulation of human presence to maintain their essential purpose. The removal of specimens is not allowed without special permission and some reserves are protected by bye-laws.

Nevertheless, the Nature Conservancy Council welcomes visitors who respect the wildlife of its reserves and who come to learn about and enjoy these fine places. The Council has established information centres at Knockan near Elphin, Beinn Eighe and Loch an Eilean near Aviemore. Other nature reserves have been established, notably by the Royal Society for the Protection of Birds, and the Scottish Wildlife Trust. Many of these have a strong botanical interest, and on some at least limited access to the public is allowed.

There is the whole of the Highlands and Islands in which to look at plants, and very few species are confined to nature reserves. There is plenty to see from almost any base the visitor cares to choose, though in this part of Britain so much depends on that uncertain and capricious factor, the weather. Much of the land is private, but except in the deer-stalking and grouse-shooting seasons, from mid August until the end of October, there are seldom real problems over access. Most estate offices and people on the ground, such as stalkers, keepers and shepherds, will readily give permission to walk on unenclosed hill land, if this is sought in a proper manner. If there is trouble, it usually arises from a few thoughtless people's failure to observe the countryside code or even the normal dictates of common-sense and courtesy. Foresters have a natural anxiety about fire hazards, particularly during dry weather, and although the Forestry Commission normally allow fairly free access on foot to their woodland tracks, restrictions may be necessary for such special reasons. Motorists should respect the prohibitions on the use of private roads, which are costly to maintain, and could, if used by all-comers, lead to an even more rapid erosion of the wilderness character of many fine areas than is already apparent.

The Highlands and Islands are marvellous country, but the scene changes steadily as the remorseless pressures of the late twentieth century spread through this once sequestered region. The earlier image of a neglected, deserted and barren land, given over largely to the seasonal sporting pleasures of wealthy outsiders, is fading rapidly. The region still contains by far the largest area of uncultivated and undeveloped land in the British Isles and so remains a paradise for naturalists. Much of it is by our standards a wilderness country, magnetic and beautiful, yet fragile and easily spoiled. Its essential character, and its richness in wild plants and animals will surely suffer steady depletion unless there are deliberate attempts to cherish and protect them. My hope is that this book may help to bring home the message that the native flora of the Highlands and Islands is indeed an asset well worth conserving, to the benefit of residents and visitors alike, and from a sense of respect and responsibility for the world of Nature in its own right.

Fig. 2 Broom in flower – a common and colourful shrub of roadsides and waste places in the lowlands.

Historical Sketch

Fifteen thousand years ago, the Highlands and Islands were in the grip of the Ice Age and the region bore some resemblance to Greenland and Spitzbergen as they are today.

On the mountains, massive snowfields lay more or less permanently over much of the high ground whilst glaciers filled the higher corries and fed the great rivers of ice gouging out the main glens below. Black bare peaks and ridges stood out here and there, free from ice and snow at least during summer, and the coastal lowlands, especially, may have held other ice-free areas, but such ground experienced a climate of Arctic severity. The limits of glaciation lay far to the south, across the English Midlands. Scotland as a whole was a barren, frozen and inhospitable land forsaken by the early Stone Age men. We shall probably never know just what kinds of plants survived during the time when ice and snow cover were greatest, or precisely where they hung on. It is most probable that some of the hardier northern types, at least, were able to live within Scotland on ground remaining free of ice. Since so much water was locked up in ice, the sea level was lower than at present, and certain coastal lands later to be drowned by rising melt-water may then have supported quite varied populations of plants, including even trees.

Eventually the climate turned for the better and slowly warmed up, so that the ice at the lower levels began to melt and retreat. The glaciers dwindled at their foot and the yearly snow cover on other ground became less. Soil and rock particles scraped along by the ice were released and built up in characteristic deposits of boulder clay or sand, either as rounded heaps, termed *moraines*, or as a more general covering of *drift*. Large fans and spreads of material were plastered over the valley bottoms as the rivers became swollen by melt water.

All this bare ground was splendid habitat for plants, which rapidly made use of the opportunities, spreading out from their nearest refuges and building up a vegetation cover that advanced steadily as more new terrain became available. Lakes of all sizes formed in the hollows and troughs which remained and these offered habitat to aquatic plants. Thus began a sequence of change which, if properly understood, can help to explain the distribution of plants as we find them today.

How fascinating it would be if we could take a time machine and travel back to view these ancient scenes in their primeval splendour, and follow the detailed migration of plant and animal life down the centuries to the dawn of our civilisation. The historically inclined botanist, sadly, has to turn to less fanciful methods of investigation.

By studying the remains of plants, such as leaves, fruits and seeds, or the stems of woody species, which were buried in lake and other deposits of sediment formed at the time, some evidence of the nature of this vegetation has been obtained. The still more important technique of pollen analysis depends on the chance that pollen grains of most plants have a relatively indestructible outer coat and are preserved almost unchanged for indefinite periods when they are incorporated in waterlogged deposits of sediment or the partly decomposed remains of plants which form peat. They can be extracted, and then identified and counted under the microscope; their relative proportions at any one level in the deposit give some indication of the composition of the surrounding vegetation at the time.

The first plants to colonise the bare debris recently left by the ice were Arctic and Alpine types such as may be found close to the glaciers and snowfields of these regions today. A low and sparse plant cover of small herbs, shrubs, mosses, liverworts and lichens advanced northwards and upwards as the ice receded, and was evidently composed mainly of species which had occupied ground just beyond the southern limits of glaciation. Where ice-free areas had persisted, any plants growing on them spread out as the area of suitable habitat expanded and eventually merged with those migrating from farther south.

On dry, windswept terrain, an open stony and gravelly 'fell-field' habitat probably persisted, with various surface patterns in the debris produced by frost-thaw movements on the ground, and a very sparse, patchy growth of plants.

On more stable ground, there developed heaths of dwarf shrubs such as crowberry and mountain avens, but containing an abundance of herbaceous plants, mosses and lichens. Where drainage was poor, in hollows and plains, a tundra-like vegetation evidently formed, with abundance of sedges, grasses, dwarf willows, dwarf birch and mosses.

Taller woody plants gradually appeared and increased, at least on damper and better soils, with medium willows and juniper in quantity. Birch was the first tree and it soon spread rapidly on all but the wettest ground, forming rather low and patchy woodland interspersed with marshes and tundra, as it does today at the limits of the Boreal forest in northern Europe—the 'taiga' zone. The spread of waves of tundra and taiga woodland suffered a temporary reversal as the climate turned colder again and the ice re-advanced down the mountains and southwards. But this set-back passed, and the warming of the climate continued again steadily up to around 5000 B.C. Scots pine began to appear and expand in Wester Ross around 6000 B.C. taking the place of birch, in the areas around Glen Torridon and Loch Maree. Pine forest developed and expanded south-eastwards, whilst the birch and fell-field—tundra zones continued to move north and upwards in their invading waves.

Only later, around 3000 B.C., did pine forest spread out more widely, reaching the Rannoch district, eastern Skye and Assynt in Sutherland, and eventually extending thence eastwards right across the Highlands to the shores of the North Sea. Scots pine was favoured especially by the large spreads of rather infertile and well-drained glacial sand occupying much of the lower ground in the eastern Highlands. More patchy areas of lime-rich soil probably continued to be occupied largely by birch wood with variable amounts of juniper, hazel and aspen. Pine seemed never to penetrate far into Sutherland or Caithness, nor more than marginally into the Hebrides, so that the extreme north and west, insofar as they were wooded at all, remained under birch, willow and juniper.

As the warming phase neared its peak around 6–5000 B.C., oak spread from the south and became dominant in many parts of the southern and western Highlands. It reached the Inner Hebrides and penetrated far north along both the east and west coasts of Sutherland, even exceeding the northern limits of pine. Oak-woods with much birch, rowan and holly were characteristic of poorer soils, but where fertility was higher there were mixtures also of ash and wych elm with much hazel as an undershrub. On the very local occurrences of limestone (*Fig. 7*), ash was dominant, and on some of the basic volcanic rocks of the Hebrides, thickets of hazel may have been the only form of wood-land cover to develop. Alder became an extremely widespread tree in most areas on the wet soils of the flatter valley bottoms, and formed swamp woods locally.

The pollen analytical studies of Drs. John and Hilary Birks and their colleagues have shown that the pattern of forest cover over the Highlands at the height of the post-glacial warm phase was substantially the same as that to be seen, albeit in a massively fragmented way, at the present day. For instance, the fine oakwood at Taynish on the Argyll coast has remained an oakwood for 7000 years, ever since this tree first became established during the far-off Boreal Period. Conversely, in the far north-east of Scotland, in Caithness, Orkney and Shetland, the evidence points to a general absence of woodland ever since the ice disappeared. Trees simply failed to establish themselves in quantity in these areas.

Some time after 4000 B.C., temperature evidently began to fall again slightly, and the present day climate is slightly cooler than during that previous period, which has accordingly been termed the Post Glacial Climatic Optimum. The difference may amount to a fall in mean annual temperature of no more than 2–3 degrees Centigrade, but even this could have had quite marked effects on vegetation. One implication is that the forest limits must have stood higher on the mountains at the Climatic Optimum than subsequently, and that there was a downwards retreat as conditions became cooler. Another is that the fell-field and tundra zone must have suffered its greatest restriction during this warmest period, being pushed farthest up the mountains by the forest and actually eliminated from lower tops which were within the tree limit; it too would expand downwards and outwards as the forest edge receded.

This is a neat theory, and one long held by historian botanists, but recent work has produced remarkably little evidence to lend it support. Perhaps the strongest indications come from the remains of well-preserved Scots pine trunks and roots of considerable size in eroding peat bogs at 1900-2000 ft., in various parts of the central Highlands. The earliest trees which grew here have been dated to around 5000 B.C. and were of a much greater stature than could ever be envisaged growing at this elevation today. The natural, climatic tree-limit (timber-line) varies from one district to another, and even within a single massif, according to differences in exposure, aspect, soil depth and so on. But the present maximum elevation for pinewood, as distinct from scattered individual pines, is at 2100 ft. on the western slopes of the Cairngorms, and this is a miserable collection of gnarled and contorted little trees. Even allowing that the upper edge of the forest has been so totally destroyed that no truly natural limit remains, it is still unlikely that pines with straight trunks up to one foot in diameter would grow now at 2000 ft. Remains of pines at still higher altitudes are remarkably infrequent, though they occur at 2600 ft. in the Cairngorms. The most important point about this theory is that, if the forest zone reached its greatest vertical extension at the Climatic Optimum, this period is likely to have seen the extinction in Britain of certain high montane plants which flourished immediately after the retreat of the ice, but later became increasingly restricted to the highest mountain tops. This could help to explain why our Scottish hills lack many plants characteristic of the Scandinavian mountains, and why many other species are highly localised within the Highlands.

Once a plant has become really rare, its powers of spread, even with the return of more favourable conditions, may be severely restricted. A population may become too small to produce the amount of seed needed to give a fair chance of overcoming the various resistances to establishment of new growth elsewhere.

There was another major change during these ancient times which is much clearer, though the underlying causes are still uncertain. This is the enormously widespread growth of peat bogs which began around 5000 B.C. and has continued ever since.

It has been presumed that rainfall must have increased at this time, leading to a general waterlogging of soils and an accompanying increase in the rate at which chemical nutrients needed by plants were washed out of the soil. These two processes, waterlogging and acidification through leaching, give conditions which resist the normal decomposition of dead plant remains, so that these tend to accumulate as a deposit of peat. At the beginning of this Atlantic Period peat began to form in many situations where before there had been well-drained soils, often with a forest cover. The bog surfaces rose vertically and spread out laterally. On the gentler moors and hills they gradually formed a complete mantle, hence their aptly descriptive name, *blanket bog*. The rate of bog growth depended on wetness, for the wetter the ground the less the plant remains decayed, and the more rapidly the

Fig. 3 Rannoch Moor, Argyll – ancient pine stumps exposed in eroding blanket bogs.

Opposite
Fig. 4 Birchwoods near Aviemore, Inverness-shire.

Fig. 5 Rowan in fruit and conifer plantations at head of Dornoch Firth.

Fig. 6 Blanket bogs with hummocks of bog moss and woolly fringe moss, near Dundonnell, Ross-shire.

living surface rose above the underlying mineral soil.

Many of these bogs are still growing actively today though careful examination of the peat shows that there has usually been periodic change in the peat-forming vegetation during the several thousand years of their existence. The peat preserves a record of the plants which grew there at each successive point in time, and in the less decomposed layers these are easily recognisable.

The various kinds of bog-moss or *Sphagnum* have been the most important peat formers, for they carpet quite large areas of bog surface and have a rapid upward growth. These plants have a rather sponge-like capacity for soaking up and holding water, and also maintain high acidity—two properties which resist their decay and which, consequently, led to their former use as antiseptic dressings. Other plants grow with their roots in the *Sphagnum* carpet and their remains later become overgrown and incorporated by the moss. The blanket bogs in many places have suffered a good deal of erosion, usually by gullying of the peat. Small bare channels and cracks appear in the bog surface, broaden, deepen and cut back to link up with each other. A network of deep gullies with intervening peat islands gradually develops as the exposed peat is loosened by wind, sun and frost, and is washed away down the gullies during rain. These channels cut down to the underlying mineral soil, and in places there is such general disintegration that sheet erosion is the final stage.

Eroding bogs typically show abundant remains of long-dead trees, usually as roots and basal stumps and less often the fallen trunks. Birch is probably the most widespread amongst these buried remains, but pine is commonly found in quantity (*Fig. 3*), and at high levels, juniper and willow are frequent. Pine stumps, in particular, can occur at all levels in the peat, but they are often right at the base. This has given rise to the theory that pine forest was widely overwhelmed and buried by the developing peat bogs at the onset of the Atlantic Period. But pine was able to grow quite widely on many of these bogs later on, and over too big a time-span to indicate a single drying phase, so that the story is probably not quite so simple. The great increase in alder during early Atlantic times (5000–4500 B.C.) supports the notion of marked increase in rainfall, for this tree favours wet soils.

Men of the Late Stone Age made their way back into Scotland as conditions became congenial once more, but for thousands of years the numbers of early man remained low and had little impact. By the time the Romans occupied southern Britain, the Highlands were a thinly populated region of pristine forest, bog, loch and river, with higher fell-field and mountain tundra zones rising in many places above the tree-line. The climate had evidently settled into its present phase, albeit with the customary ups and downs of temperature and rainfall which we have known since records were kept. It was nevertheless evidently a somewhat inhospitable environment, and human settlement and influence did not become marked until centuries later. Some of the earliest signs of man's presence were at intermediate levels, where the dense forests of the valleys thinned out sufficiently for the first settlers to make clearings and graze their animals. The early hunters, stock raisers and then nomadic pastoralists were replaced by the cultivators, who stayed put as the farmers of better lands, first in the east of the Highlands. The pollen record in the peat bogs shows their influence in an increase in abundance of heather, as forest was cleared and replaced by open moorland, but even more in the presence of the pollen of grasses and common weeds of cultivation, as the ground was ploughed for crop-growing.

The evidence of widespread destruction of the primitive forests steadily increased, to the point where historical records took over. From the beginning of Christian times, right up to the early nineteenth century, there was a continual onslaught on the woodlands. By the time of the Norman Conquest, the Highlands were probably still largely a primitive wilderness, with vast tracts of forestland. Deforestation reached its height during the Middle Ages, when wood was in such demand as a building material and a fuel. In the 17th and

Fig. 7 Ashwood with primroses and mossy rocks, Rassal, Loch Kishorn, Ross-shire.

18th centuries, large amounts of pine were sent from the forests of the Spey and Dee valleys to Inverness and Aberdeen, many of the logs being floated down the rivers by way of transportation.

Perhaps the most extravagant use of timber was in the smelting of fuel. The name 'Furnace' here and there points to former bloomeries where local iron ore was processed but much of the timber went to smelt-mills farther south. Clearance for grazing and arable land continued apace, and it is said that large areas of forestland were burned simply to remove refuges for wolves and other undesirable creatures. While there was some effort at replanting in places, it was far outweighed by the rate of loss. The large numbers of grazing animals in many areas severely limited natural regeneration by cropping down the seedlings.

By 1800 the appearance of the Highlands was transformed. Arable land was restricted mainly to lower levels and to the drier east, and the great expanses of mountain and moorland were used as grazings for large numbers of cattle, goats and sheep. Then, from 1810–1830, came the grim episode of the Highland Clearances, a period of enforced human depopulation over much of the region, followed by a phase of intensified sheep rearing in the uplands. This gave way to the great Victorian burgeoning of sporting estates, given over to management for the native red deer and red grouse. Interest in sheep became more localised, mainly in the west and on the most fertile hills. All three forms of land use involved deliberate burning of the vegetation as a management technique, to remove the old, coarse material and encourage a new and more nutritious growth.

For decades there was little change. Then, in 1919, the Forestry Commission was created and began to re-afforest the lower hill slopes and moorlands in many places, first using mainly Scots pine and larch, but later turning increasingly to Sitka spruce and lodgepole

Fig. 8 Scots pine forest at its climatic limit at 2,100 ft., on Creag Fhiachlach, Cairngorms, Inverness-shire.

pine. Private landowners and forestry concerns have also done a great deal of tree planting, and throughout the Highlands there has been a great increase of woodland cover, though this is nearly all of a particularly artificial kind, with a flora usually notable for its poverty. Sheep numbers have increased on many deer forests and grouse moors, and the present pressures to utilise land to its maximum productive capacity have led to an increase in improvement of marginal hill pastures, reclamation of moorland and an increased campaign of bracken eradication.

Other fairly recent changes have been the development of hydro-electric schemes, which have transformed the appearance of many glens and mountains; the encouragement of tourism and recreation, expressed particularly in creation of ski-ing and other holiday facilities at Aviemore and on the Cairngorms; and finally, the coming of industry to the Highlands.

The Age of North Sea Oil is upon us, with all its implications for good and ill, and the prospects for other forms of mineral exploitation on land are being carefully assessed. Hand in hand with all such modern developments go the need for improved access and transportation. Road improvement schemes have blossomed and remote country becomes ever easier to reach. One notes with some sadness that so many landowners have found it necessary to drive Landrover and tractor roads over wild mountains and moorlands, but this is yet another sign of the times. All these changes have effects on vegetation, ranging from the direct and obvious to the indirect and subtle. The story of the Highlands since the Ice Age is one of continual change, but with man increasingly replacing climate as the major cause, during the last 2000 years.

Fig. 9 Deserted croft beside Berriedale Water, under Morven, Caithness.

An Ecological Background

The Highlands and Islands are the biggest tract of mountainous country in the British Isles, containing the highest peaks, and formed of the oldest rocks. To the geologist and geographer, the region is defined by the Highland Boundary Fault, a massive ancient dislocation running diagonally north-eastwards from the Isle of Arran in the Clyde, through the south end of Loch Lomond to Stonehaven on the east coast. Defined administratively as the Highland Region, the Argyll and Bute district of Strathclyde, the Western Isles, Orkney, Shetland and the Island of Arran they form the territory which is the responsibility of the Highlands and Islands Development Board and therefore the subject of this book. The botanically important counties of Perth, Angus and Aberdeen are omitted, though some reference will be made to them. To the naturalist—and the wildlife—political boundaries are usually highly artificial, and most people will be more concerned to appreciate the Highlands as a natural region.

Within the Highlands and Islands there are important differences in conditions which have a profound influence on the distribution and abundance of plants. Some understanding of these relationships will help to enrich the interest of looking at plants in the wild. Each plant or piece of vegetation has a story to tell, if we can but read the signs. Within the seemingly endless variety and complexity there is some kind of order and pattern, and it is interesting to be able to work this out for oneself. In the same way that the good gardener develops an awareness of the varying needs of the different plants he grows, the experienced field botanist acquires an eye for habitat. He remembers the conditions under which each species of plant grows, the other species with which it is associated and the repetitions and variations in these combination. A memory of pattern builds up almost automatically, and it soon becomes apparent that certain plants of limited distribution indicate certain conditions, whether of climate, soil or human influence. These indicators often help to focus the search for other species, especially the rarer kinds.

This sort of ecological awareness grows out of nothing more than intelligent observation and a sense of curiosity. It allows plant-hunting to be far more purposeful than mere random search, and gives the satisfaction of finding plants in places where they had hopefully been expected.

Human beings are familiar with the capriciousness of the weather in this country, but it is average climate which is more important to the plants. We have seen how important a part changing climate has played in moulding the character of the vegetation over many thousands of years. The present climatic conditions and their pattern of variation remain one of the major influences on the flora of the Highlands today. Mountaineers are only too well aware that conditions on the high tops tend to be very different from those in the valleys, but climate also changes markedly in general character as one travels across the Highlands, especially from east to west.

If we look first at the purely local variations in average climate, it is fairly obvious that these are related especially to altitude. The higher one goes the greater is the rainfall, cloud cover, snow cover and windspeed, and the lower the temperature and amount of sunshine. Steep north to east slopes are shadier and therefore damper than those more exposed to the sun. The higher mountains often show a quite well-marked zoning of vegetation with increasing elevation.

Farmland often occupies the valley floors and lower slopes, but within the range of natural and semi-natural vegetation, forest is the lowest of these zones on drier ground, and reaches its highest level of just over 2000 ft. in the Cairngorms. Woodland has, however, been extensively replaced by heath, grassland and bracken as a result of deforestation, and seldom shows a true upper limit. Originally a fringe of scrub about 3–4 feet high girdled the upper edge of the forest, and consisted mainly of juniper and willows. This scrub zone, which has been destroyed almost totally by fire and grazing, gave way still higher up to much lower growths of dwarf juniper, heather and other dwarf shrubs forming alpine heaths. On the windswept upper spurs and summits these shrubs become completely flattened on the ground, and are finally replaced at over 3000 ft. by dense carpets of moss, lichen or small grasses and their relatives, though on many tops there is a great deal of bare soil and stones, kept in this state by the action of frost and wind. This kind of terrain is called 'fell-field' and is a habitat characteristic of the high Arctic regions close to the edge of permanent snow fields and glaciers. It is one of the harshest of all environments for living creatures.

Fig. 10 A common woodland moss (Thuidium tamariscinum).

Varying snow-cover and drainage modify the simple pattern of zonation. Sheltered hollows and shaded slopes where patches of snow lie late into the spring or even summer show change from prevalence of heather to that of bilberry, or crowberry on rockier ground. Still later snow areas have dense beds of mat grass, and under the longest-lasting patches of all, melting out only well into the summer, there is merely a carpet of mosses, lichens and a sparse growth of small alpines. At levels above 3000 ft, most of the vegetation is influenced by late snow cover in some degree, except on exposed ground where severe winds prevent the build up of snow. The steady melting of these snow patches tends to produce a good deal of wet ground below, and there are many cold springs, flushes and rills in the high corries, usually with cushions of moss and liverwort, and many alpine plants. These wet habitats also occur widely on the hills away from snow patches, and where the seepage water spreads out through the soil there are little marshes, often with low growths of sedge or rush. These marshes are quite extensive on many lower slopes and in the valley bottoms, or around shallow lake edges.

The blanket bogs (*Fig. 6*) mentioned earlier occur on poorly drained flats and gentle slopes at all levels up to 3300 ft., and their living surfaces apparently depend for water and mineral nutrients entirely on rainwater and dust from the atmosphere. They are composed mostly of moisture-loving plants such as bog-mosses, sheathing and common cotton grass, deer sedge, purple moor grass, heather, cross-leaved heath, bog asphodel and sundew.

At the other extreme, the steeper mountain sides break out into abrupt cliffs on which soil occurs only on ledges or in crevices, and where gravity maintains a permanent instability. These precipices are often extremely productive in alpine plants, since they are free there from both the competition of more robust vegetation and the depredations of grazing animals. Litters of bare rock debris or scree cover many slopes and summits, and show all degrees of colonisation by plants, especially ferns, mosses, liverworts and lichens.

These are, in briefest outline, the main kinds of vegetation to be seen in a typical mountain area. Coastal areas have distinctive maritime habits such as sea cliffs, rocky shores, shingle beaches, sand dunes and, very locally in the Highlands, salt marshes. Anyone travelling widely around the Highlands and Islands, however, will soon become aware that, from place to place, there are some quite marked differences in the plant life of comparable situations. The reasons are various, but differences in climate between separate districts are one of the most important. The west of the Highlands and the Western Isles are an extremely humid but mild district. The prevailing westerly winds across the Atlantic shed large amounts of rain when they are forced upwards by the high mountain ranges. Even more important, though, is the number of days per year on which measurable rain falls: it exceeds 250 in many areas, and results in a constantly moist atmosphere, which in turn produces a general prevalence of sour, peaty soils. Blanket bog is extensive and has formed on steeper slopes than in the east, whilst there is a great extent of damp heath or grassland on the drift-covered lower slopes. Moisture-loving plants flourish and on some of the shady north to east facing mountain slopes are luxuriant carpets and cushions of leafy liverworts of a type found only in the most humid regions of the world.

The Atlantic currents also have a warming influence along the whole western seaboard and this extends even to Shetland. The winters do not give extremes of cold, as is shown by the cultivation of early spring flowers on the Isle of Tiree, and the richness of the famous gardens at Inverewe on the coast of Wester Ross. It is also reflected within the native flora by the presence in the more southerly parts of the region of plants which cannot withstand cold winters. Whilst the winters are much less severe than those of the eastern Highlands, the summers are distinctly cooler. Extreme windiness is another feature of the oceanic climate, and the western coast and its adjoining mountain ranges are at times blasted by ferocious gales, especially during winter. There is also a decrease in average temperature as one goes north in the Highlands which superimposes another gradient on that found between west and east.

There is thus a contrast between the relatively continental conditions of the eastern Highlands and the extreme oceanic climate of the west.

Even in the east, closeness to the sea has the effect of raising winter temperatures. The coast from Nairn to Elgin on the south side of the Moray Firth probably has one of the most congenial climates in Britain; it is dry, sunny and sheltered, with pleasant warm summers but lacks the extremes of winter cold which characterise inland eastern districts, and avoids also the unpleasant fine weather sea mists that plague so much of the east coast of Britain.

The various meteorological conditions interact, with the result that the overall climate becomes increasingly unfavourable for plant growth in a north-westerly direction. Increasing windiness, humidity and lack of summer warmth are the factors especially responsible, and their effects are shown most obviously in a downward shift of the upper or lower limits of the natural vegetation zones. Although it is represented only by fragments, the upper limit of woodland falls from over 2000 ft., in the Cairngorms to about 1000 ft. in north-west Sutherland. The lower limits of alpine heaths dominated by dwarf shrubs drop even more, from about 2500 ft. to only 1000 ft., and on the most severely storm-swept coasts of the north-west, these communities descend virtually to sea level, since no competitors in the form of taller woody plants can withstand such severe conditions. Many individual kinds of plant in the far north grow fully 1000 ft. below their lowest occurrences in the central mountains.

Climate has a powerful influence on land use and hence on the degree to which the native flora has been modified. The most favourable areas for farming are in the lowlands of the east, where the amount of warmth and sunshine give a long growing season, and dryness reduces soil leaching and peat formation to lower levels than in the west. The coastal lands from the Firth of Forth to the Black Isle are an important agricultural belt, and it is noticeable that in counties such as Nairn, Moray, Banff and Aberdeen, arable farmland ascends the lower moorlands to fully 1000 ft. In many places here the heather moors abut sharply on fields of oats and barley, to which the grouse descend for stubble gleaning in the harder back-end of the year. Root crops do well on the upland farms and there are rich grasslands on which cattle are fattened.

A journey to the west coast shows a complete contrast, though in part this results from differences in geology and land form. In many places there is no room for arable farming, for the mountains plunge steeply to the water's edge in fjord-like sea lochs. Even where the coastal lands are low, they are often rugged, with too much bare rock or open water to allow attempts at farming.

The general scarcity of good soils is another natural obstacle to cultivation, and the prevailing peaty soils can carry only low numbers of grazing animals. Only here and there, in favourable situations and often through sheer determination, has land been won for the growing of crops. In places the wind has spread large amounts of shell sand onto the shore, or the main rivers have built up fertile deposits of sediment along their less turbulent lower reaches (*Fig. 9*).

The desperation of evicted crofters, dumped in forlorn outposts during the Clearances, created patches of potatoes and oats, hay meadows and pastures, amongst the bogs and rocks, through the assiduous carting of the natural fertilisers, seaweed and sand, from the shore. In a very few places, nature was kinder, in providing exposures of limestone which could be exploited with better results, as may be seen around Elphin and Durness in Sutherland today, where the prevailing greenness of the scene is in marked contrast to the drab sterility of the rocky moorlands surrounding these oases.

Fig. 11 A common liverwort of moist, shady places (Conocephalum conicum), bearing fruiting bodies.

The nature of the rock is always another factor of great importance, in its effects on both the kinds of land form which have developed over the ages, and on the fertility of the soils providing plants with their anchorage and nourishment. The feature of special significance to soil formation is the amount of lime (calcium carbonate) in the original rock. Lime-rich rocks tend to form fertile soils free from raw undecomposed peat, supporting a large number of different kinds of plants in any one place. Rocks poor in lime give rise to soils which are mostly sour and acidic, with a strong tendency towards raw peat development at the surface, and their flora is usually rather limited in variety.

Washing out from the soil of nutrients needed by plants is one of the main soil forming tendencies throughout the Highlands, especially in the west, where so much rain falls. This process of leaching results in a net loss of essential elements, particularly over rocks which are hard and poor in these substances anyway, and it goes hand in hand with the growth of acidic peat in waterlogged situations. By contrast, on limestone or other calcareous rock, nutrient levels are more often maintained, and the water draining from such materials is especially effective in enriching the ground which lies in its path, in the process known as flushing. Lime-rich rocks are, however, very local in the Highlands, and there is a general prevalence of acidic types. The most important calcareous mountains actually lie outside our area, in Perthshire, but most districts of the Highlands have what the botanist regards as 'good' rocks with a distinctive richness of flora. It is particularly useful to be able to recognise some of the plants which indicate soils or water with a high lime content; they are often the pointer to especially choice habitats with an assortment of good things.

Rock type noticeably affects the distribution of native trees, where these still remain. Woods of sessile oak and Scots pine in pure stands are characteristic of poor rocks and soils, whereas mixed deciduous woods with oak, ash, wych elm and hazel occur on better ground, and ash with hazel is usual on limestone. Birch (*Fig. 4*) covers the whole range of soil types, but is often a replacement after other woodland has been cleared. Where woods on fertile soils are not grazed by large animals, they usually have an abundance of undershrubs and a herbaceous vegetation in which colourful and conspicuous flowering plants predominate. When sheep or deer have free access there are nearly always fewer shrubs and a prevalence of grasses. Ungrazed woods on infertile soils also have undershrubs, but dwarf shrubs such as heather and bilberry are more prominent than herbs, except perhaps the great woodrush. Grazing again reduces the underscrub and causes grasses to flourish.

Geology does indeed have as profound an influence on the development of agriculture as does climate. Both wild and domesticated animals have an instinctive preference for the vegetation of the richer soils, and graze this selectively. Early Man also appears to have had a primitive awareness of the advantage of exploiting first the most productive soils, and this choice has been increasingly reinforced as farming has developed.

Arable farming depends on the occurrence of fertile soils, and the exploitation of the uplands as grazing land has also been heaviest in the areas of lime-rich rocks and soils. The densest sheep numbers are on the calcareous mountains and this is where modification of the original vegetation has been greatest. Grasslands prevail in such areas and there has been considerable loss of shrub vegetation. In any area, heavy grazing, whether by sheep, cattle, goats or deer, and accompanied by the inevitable burning of vegetation, has had the effect of eradicating dwarf shrubs, especially heather, and causing their replacement by grasses or their allies. In parts of the more southerly mountains, especially, grasslands predominate and heather is somewhat local, often showing restriction to crags or steep rocky slopes where it has some protection from fire and sheep.

On dry ground, heather is replaced by bent grass and fescue, but as the soils become moister, mat grass, heath rush, deer sedge, purple moor grass and cotton grass are the species which take possession. These plants were usually present already, but they are less affected by grazing and burning than heather, and so can spread out as it retreats.

The dry grasslands are often extensively invaded by bracken, which is at an even greater advantage, for it is little grazed, and its underground rhizomes are unaffected by burning of the surface vegetation, which is normally done early in the year before the young fronds appear. In some drier districts bilberry sometimes replaces heather as an intermediate stage before being itself ousted by grasses. Indiscriminate burning seems to have particularly severe effects in the western Highlands, especially in reducing the abundance of certain alpine dwarf shrubs, such as dwarf juniper, common bearberry, alpine bearberry and dwarf birch.

Fire and high numbers of animals in combination, can be a potent cause of soil erosion, especially on steep ground. Many of the loose expanses of rock litter known as scree have formed in this way, though some have resulted more directly from deforestation. The extremely widespread erosion of blanket bogs by gullying from the edges inwards also seems to be connected with these influences. Some ecologists believe that under the heavy rainfall of the west this extractive use of the hill land has also resulted in a substantial loss of soil fertility. The Highland laird, Osgood Mackenzie (1842–1925) who founded the celebrated Inverewe gardens, observed during his long life permanent declines in many kinds of birds, other animals and plants which seemed to reflect such a reduction in carrying capacity. Some of the more barren parts of these western uplands now have a productivity akin to that of semi-deserts, and their flora is extremely limited.

In the drier, eastern parts of the Highlands, especially on the more gently sloping hills where red grouse could be driven over guns, large areas of moorland have long been managed mainly for this bird and the sport it provides.

Heather flourishes on these dry hills and has been managed by a careful rotational system of burning during the winter and early spring. Different patches are fired each year, so that the moor develops a mosaic pattern of varying colour tones representing stages in the regrowth of the heather. The young plants have a higher nutritional value than the old leggy growths and thus enhance the carrying capacity of the moor. The numbers of sheep are deliberately kept low so that they do not suppress the heather or compete unduly with the grouse.

Deer forests occur all over the Highlands, but especially on the higher mountain ranges where the rocks are hard and acidic, for the red deer is able to live quite well on ground where sheep need supplementary feeding, especially in winter. Most deer forests carry some sheep, and they tend to be managed in the same way as sheep walks, with rather haphazard burning, which has locally reduced the dwarf shrub cover.

Cattle have recently regained some popularity in certain upland areas. They are less choosey grazers than sheep and will crop down coarse grasses but, from their size and weight, they have a more marked influence on the ground through their treading and manuring.

All these managed herbivores live predominantly within the levels once occupied by forest and scrub, so that, except on the wettest

Fig. 12 Purple oxytropis on mountain rocks, with alpine ladies' mantle.

Fig. 13 Wood vetch on woodland floor.

Fig. 14 Water lobelia, in shallow water at loch edge.

Fig. 15 Beech Fern, a widespread fern of woods and shady rocks.

Fig. 16 Orthothecium rufescens – one of our most beautiful mosses, growing on basic mountain cliffs.

Fig. 17 Globe flower – a characteristic northern plant of ungrazed meadows, river banks and cliff ledges.

ground, the vegetation of their habitat is mostly of a secondary kind.

Whilst heather was evidently plentiful in natural gaps amongst the trees in the original forests, the great sweeps of heather moor for which the Highlands are famous are almost wholly artificial.

The use of the uplands as grazing land has seen a severe decline in many plants which were once abundant. The mountain willows and many tall herbs and alpines in particular have suffered great contraction in distribution, and some species are more or less confined to inaccessible rock ledges. The ungrazed hay meadows are mostly less interesting and attractive in their wild flowers than in former years, as a result of modern farming techniques, which involve 'improving' such grasslands by adding fertilisers or herbicides, or even ploughing and re-seeding with commercial grass strains.

The great reforestation programmes of the last fifty years have restored woodland (*Fig. 5*) on the lower hill slopes over large areas, spread right through the Highlands. Yet these new woodlands have little interest for the botanist. They are nearly all of alien conifers, usually grown in such dense stands that the intense shade and deep needle litter exclude all but a few of the common ferns, mosses and toadstools. Some of the densest woods of Sitka spruce have a forest floor almost devoid of plant life. Only along the rides, tracks, or streamsides does other vegetation appear again.

There is nevertheless plenty left to see, and the Highlands offer perhaps greater diversity of habitat and flora than any other part of Britain. The effects of human activity are not wholly on the debit side for the botanist for, whilst some plants have declined, others have increased and spread in response to the creation of new habitats. Some kinds of terrain remain more or less inviolate, defying human intrusion to cause change. Among these are the rocky coasts and mountain crags, the high alpine summits, and the remote hill lochs. Here there remain the last refuges in this country of vegetation which is truly natural in the sense of being virtually unchanged by the hand of man.

Fig. 18 The high peaks and corries of the Cairngorms, on Braeriach, Inverness-shire, showing late snow patches and corrie lochans.

Highland Botanists

An account of the plant life of the Highlands should pay some tribute to the work of the many botanists whose labours have gradually built up the massive fund of knowledge available to us today. Dr. H. R. Fletcher, formerly Regius Keeper of the Royal Botanic Garden in Edinburgh, has written a detailed historical account of the subject which should be read by anyone interested in learning the full story. I have drawn heavily from it in making this summary.

There is space here to mention only some of the more notable figures, at the risk of being unfair to the rest. The first published account of Scottish plants was in a natural history of Scotland by Sir Robert Sibbald (after whom the little alpine Sibbaldia was named) in 1684, and this listed nearly 500 different kinds.

For nearly a hundred years only isolated new records of Scottish plants appeared, but in 1778 the Rev. John Lightfoot produced his *Flora Scotica*. Lightfoot journeyed through the Highlands in 1772 with Thomas Pennant, correspondent of Gilbert White, and the two recorded plants as they went, finding several new to Britain. The *Flora* listed 1250 species of flowering and lower plants, including over 50 alpines, and drew attention for the first time to the tremendous richness of the Breadalbane range, with Ben Lawers and its neighbours. The information about this splendid area came mainly from the searches of the Rev. John Stuart of Killin, who was especially helpful to Lightfoot. Much information also came from John Hope, Professor of Botany in Edinburgh, who with his students was active in many parts of the Highlands and several of the large islands.

Probably the greatest of these pioneers, in the extent of his explorations and his unremitting energy, was George Don, a nurseryman of Forfar. Don put the second great area for alpine plants, the Clova-Lochnagar mountains, firmly on the botanical map, but hunted other major massifs—the Cairngorms, the Lawers range, Ben Nevis, the Loch Lomond hills, and the remote country of Knoydart. He worked between about 1779 and 1810, and added many new plants to the British list.

Don fell under some slight suspicion, for he cultivated a great many plants, both native and alien, in his garden, and it was surmised that he might have planted some introduced species in the hills, or become confused about the origin of certain others which he claimed as natives. However, most of Don's discoveries are undoubtedly native, and several which were long-lost have been re-discovered one by one. Whilst he may have made a few genuine errors over his cultivated material, George Don seems to have been a remarkably fine and discerning field botanist, who probably added more to knowledge of the Highland flora than any other person, before or since. His most important publication was an account of the plants of Forfar (Angus) in 1812, though his records in general were incorporated in the compilations of other botanists. Another nurseryman, James Dickson, made several important finds on Ben Lawers during the same period.

Two leading figures in British botany were themselves active in the Highlands during the first half of the 19th century: Sir William Hooker, the systematist, who wrote the British Flora in 1830, and Hewett C. Watson, the plant geographer, whose series of works from 1847 onwards on the distribution of our native plants broke new ground.

More and more plant seekers journeyed every year to the Highlands, urged on by the growing awareness of the richness of the region in alpine and northern species. New species continued to be found, and there was a steady expansion in knowledge of the distribution pattern of each plant.

In those days professors of botany saw to it that their students were well acquainted with whole plants. The redoubtable Professor John Hutton Balfour of Edinburgh, reputed to be something of a martinet in the best style of Victorian academics, forayed annually into the mountains with his parties of students. On one occasion, believing themselves to be on a right-of-way, the party found their path barred by the retainers of the Duke of Atholl. The dispute was taken up in court and became commemorated in verse as the 'Battle o' Glen Tilt'.

The Aberdeen Professor, William MacGillivray, was another indefatigable hill traveller, and the fruits of his explorations of the Cairngorms and their corries are carefully presented in detailed lists in his remarkable book *The Natural History of Deeside and Braemar*, published posthumously in 1855. MacGillivray had a distinctly ecological approach to his survey, listing groups of plants according to altitude and major habitats.

Among the many visitors from farther afield were the two James Backhouse, father and son, Quaker botanists from York and also nurserymen, who came nearly every year over the period 1845–1861. They were remarkable finders of plants, missing practically nothing and never making a mistake. The younger Backhouse in particular had a nose for rare plants and gained the reputation in his day of having the foremost knowledge in Britain of the Scottish mountain flora. He was also the first botanist to get to grips with that exceptionally difficult group of plants, the hawkweeds (*Hieracium*), and was in advance of many modern plant taxonomists in that he grew all his plants to find out which characters remained the most constant. Not surprisingly, most of his hawkweed species still stand to this day, though the total number is now vastly increased and intelligible only to a specialist in the group.

Around this time, botanical clubs and natural history societies began to appear and flourish, and within Scotland there were notably active groups in Perth, Edinburgh and Glasgow. The Scottish Alpine Botanical Club was founded in 1870, its original members including Hutton Balfour. These societies used to hold excursions into the mountains and elsewhere, and accounts of their finds, often fascinating to read and with valuable lists of plants, appear in their own journals and other botanical periodicals of the time. Individual explorers also contributed their own descriptions of field trips to the literature, and these make interesting reading too.

County floras began to appear, though at rather long intervals. William Gardiner, a Dundee umbrella maker, published his Flora of Forfarshire in 1848; George Dickie, Professor at Aberdeen, contributed the Botanist's Guide to the Counties of Aberdeen, Banff and Kincardineshire in 1860; and Francis Buchanan White, a fine all-round naturalist, worked on a Flora of Perthshire which was published in 1898, four years after his death.

The achievements of these early botanists are the more remarkable when one considers the physical difficulties which they had to overcome. They worked before the age of tarmacadam and the internal combustion engine made it all so much easier, and in their day, even reaching the foot of the hills could be a major undertaking. They journeyed by horse-drawn coach or on horse-back for as far as they could, and then walked. Many roads which we take so much for granted nowadays simply did not exist then, and the distances covered on foot were often considerable. Some of the pioneers were tremendous walkers, and their written accounts sometimes casually hint at the marathon routes and the length of the days spent in the field.

Some of the country inns were evidently more accommodating about mealtimes in those days.

And although mountaineering was only just beginning to emerge as a skilled sport by the end of the 19th century, some of the alpine plant hunters were clearly at home on the crags. Their footwear and clothing would doubtless be frowned upon by modern climbers, but they managed to explore some quite difficult ground.

The tin collecting box or vasculum was standard equipment, for they were collectors to a man, but they were making knowledge, and the material they took of first discoveries was allowable. The good areas soon became famous, though, and their riches were plundered by a steady succession of eager collectors. Only the inaccessible nature of their cliff habitats prevented some alpines from being eradicated altogether in certain places.

The present century has seen a great expansion of knowledge of plant distribution in the Highlands and Islands, as more and more botanists appeared on the scene. There is space to mention only a few, but one of the most devoted was the indefatigable Claridge Druce, who worked so industriously on the distribution of British plants, and spent much time in Scotland. Specialists came to study the 'critical' groups of hawkweeds, eyebrights, ladies' mantles, and others, and the lower plants, especially mosses, liverworts and lichens, received increasing attention. There was a beginning to the study of vegetation, as distinct from individual species, and this was exemplified by the mapping surveys of the brothers R. and W. G. Smith, and by the detailed work of the geologist-botanist C. B. Crampton in Caithness. Some of the more recent books which deal with the Highland flora and plant communities are listed in the selected bibliography.

The long period of study of the distribution of the different flowering plants, ferns and their relatives culminated in 1962 in the publication of the Atlas of the British Flora. This magnificent compilation by Dr. F. H. Perring and Dr. S. M. Walters for the Botanical Society of the British Isles was made possible by the dedicated efforts of several hundred botanists in recording the occurrence of each species in the ten kilometre squares of the National Grid over a six year period, and in assimilating all the older records made by earlier generations of plant hunters. It is a tribute to the total collective endeavour since the very first records were made. And the search still goes on to such good purpose that a revised edition of the Atlas with up-dated maps of the rarer species appeared in 1976. Further revisions of the whole work are planned, and mapping of the main groups of lower plants is proceeding apace.

Exciting finds continue to be made, with several rare alpines found in new localities during this last summer. The possibility of finding a species previously unknown in Britain is quite strong, and gives extra zest to the search. There are also several long-lost plants which could yet be re-found. While travel in the Highlands and Islands is relatively easy these days, with good roads and fast cars, and a great deal can be seen with little demand on the personal energies, the area still offers much to those with the urge to explore and the confidence to rely on their own two feet as a means of transport when the tarmac ends. There is still much wild and remote country, with many an unknown glen, crag or summit, where the modern plant seeker can feel on more equal terms with the early pioneers.

Fig. 19 Alpine ladies mantle, a common mountain plant.

Fig. 20 Glaucous meadow grass, with roseroot and woolly fringe moss, in a high rock crevice.

The South-West Highlands

The southernmost point of the Highlands and Islands is the Mull of Kintyre, lying well to the south of Glasgow. The county of Argyll, to which it belongs, is remarkable for its range of physical conditions and thus for its variety of wild plants. There is a long and indented coastline, with sheltered arms of the sea running deeply into the hills bounding the north side of the Firth of Clyde. In the middle of the Clyde itself is the large island of Arran, belonging to the county of Bute. This is the mildest part of the Highlands, and in places close to the sea there is very little frost in winter. It is a humid district, but less so in Kintyre than in the more mountainous country farther north.

Let us assume that we enter the area via the Loch Lomond or Loch Long road to Arrochar. There is then the choice of touring the steep-sided glens and ribbon-like lochs of the Cowal peninsula, or heading through Inveraray and Lochgilphead to Knapdale and Kintyre. It will soon be obvious that, apart from Loch Eck, these lochs are inlets of the sea, for there is a tidal shore, which is mostly rather narrow and rocky, with masses of olive-brown and green seaweeds festooned over the boulders. At low water, their glistening, slippery growths form a distinct zone around the loch edges. Here and there, usually around the outflow of rivers into the head of the sea lochs, sediment has built up and been colonised by plants to form fringes or patches of saltmarsh.

Thrift and scurvy grass are amongst the most conspicuous plants here, but grasses usually predominate. Within the sheltered confines of these sea lochs, where there is little wave and spray action, woodland extends virtually to the sea shore in many places, and is a most important and convenient botanical habitat within which to begin our explorations.

The lower slopes of the hills of this district are quite well wooded and though much of the woodland is coniferous and artificial, there are many patches and quite a few large areas composed of the native trees, especially oak, ash, birch, alder, and hazel. Frequently these native woods have a rocky floor, thickly strewn with blocks and variously broken by outcropping bed rock. The innumerable fast-running streams draining from the mountains have often cut deep ravines along their lower courses. These are typically grown with at least a fringe of native woodland, for their precipitous character has discouraged any attempt to extract timber, whilst some lie within bigger woods. Many such glens have picturesque waterfalls and cascades, flanked by abrupt crags and tumbled rocks, and some of them are dark, shut-in places. These rocky woods and dank ravines are fascinating places to explore for plants, though many stream gorges are dangerous. Some are best left alone (*Fig. 54*).

Oak and birch are the principal trees of the poorer soils, with rowan and holly as their most usual companions. The richer soils may have these trees, too, but ash and wych elm are usually in quantity, and hazel is the characteristic undershrub. Other trees and tall shrubs in much lesser and patchy abundance here are gean or wild cherry, aspen, hawthorn and blackthorn. Alder is usually associated with wetter soils, and is sometimes accompanied by willows. The woods here lack the variety of trees and shrubs so usual in southern England, but in a rocky glen on Arran, Glen Dhiomhan, a fringe of woodland has an abundance of two small trees, related closely to the mountain ash and the whitebeams, which are known nowhere else in the British Isles nor even in the world.

The oak and birch woods have a rather limited assortment of flowering plants on the ground beneath their canopy, and the common grasses, such as bent, sheep's fescue, wavy hair, vernal, soft fog, Yorkshire fog, and purple moor grass are amongst the most prevalent kinds. Grazed down bilberry is usually plentiful and there are tormentil, heath bedstraw, wood sage, cow-wheat, golden rod and foxglove. The last, carrying tall spikes of large pink-red bells with spotted throats, is surely one of our most handsome native plants (*Fig. 28*). How much more the foxglove would be appreciated if it were rare instead of common. For it is one of our most abundant large herbs, flourishing on the recently disturbed soils beside new roads and conifer forest tracks. This is a highly poisonous plant from which the alkaloid drug digitalin can be extracted.

Where the soils are not too acidic, there is often an abundance of wood anemone, wood sorrel, greater stitchwort, earthnut and bluebell. In Scotland the bluebell is not confined to woods, but occurs in many quite open situations on the lower hillsides, and on steep slopes above the sea. The wonderful sheets of blue which it produces at the end of May in many places excite the admiration of continental botanists, for it is a very local plant in Europe, and found mainly along the Atlantic seaboard. It is one of the lovely late spring flowers which, like the foxglove, would be still more cherished if it were rare.

Some hill woods have a good deal of bramble, but this is more typical of ungrazed lowland woods. The great woodrush and honeysuckle are also noticeably more profuse in woods not open to sheep, or on the ledges of rock-walled ravines. On the still more fertile rocks and soils, where ash, wych elm and hazel largely replace oak and birch, there is a greater variety of herbaceous plants. Dog's mercury and wild garlic are two of the most characteristic, and the predominant grasses are species such as wood falsebrome, wood melic, cock's foot and tufted hair grass.

The other common herbs include enchanter's nightshade, sanicle, herb robert, heath violet, primrose, bugle, wild strawberry, barren strawberry, lesser celandine, herb bennet, wood pimpernel, self-heal, woodruff, and wood sedge. In sheep-grazed woods, many of these plants are at their best on rock outcrops or ravine sides where they are more protected from the relentless nibblers. The tutsan is one of the less common herbs of the rocky glens, and others include wood cranesbill and wood vetch.

The ivy is less in evidence as a climber on the trees here, and grows more commonly on rock faces, for it also seems to be relished by sheep. The common polypody fern is the most typical of the higher plants actually growing on the trees themselves.

Botanists accustomed to drier districts will be struck by the contribution which the lower groups of plants, the ferns, mosses, liverworts and lichens, make to the vegetation of these western woods. These plants flourish everywhere, especially in the rockier woods with outcropping faces and block litters, but most of all in the shady, humid depths of the waterfall ravines.

Many of the more common ferns grow to a great size and to venture amongst them soon after rain is to invite a soaking. Bracken is everywhere, but in varying quantity, and there is usually abundance of the various forms of male fern, the lady fern, broad buckler fern, mountain fern, and hard fern. The splendid royal fern, which can grow up to 8 feet tall, or even more, was once common in such places, but the craze for fern collecting and growing which blossomed during the second half of the nineteenth century depleted its populations greatly. While still widespread in this district, the royal fern is now often found mainly on steep rock ledges, both inland and by the sea. Among the more delicate species, the beech fern and oak fern often grow in distinct patches or colonies, their soft and fragile fronds thriving best in rocky places where they are not trampled by sheep or deer.

These are all ferns of the more acidic rocks. Many ravines are cut along or across geological faults, and often show a change from acidic rocks to types containing a good deal of lime. The change is marked by the presence of the graceful prickly shield fern, the strap-like fronds of hart's tongue, and rosettes of the maidenhair spleenwort, whilst shady crevices and overhangs have delicate tufts of brittle bladder fern. The black spleenwort often grows here, but is one fern which seems to do better in more exposed and sunlit situations.

In really damp and shady places, especially in north-facing woods or deep ravines, there is an intriguing little fern which looks more like a moss. This is Wilson's filmy fern, whose tender, dark green and translucent fronds are only a couple of inches in length, but arise thickly from a network of wiry little stems and so form matted patches on the rocks or tree bases. Where spray from waterfalls produces a constant mist to moisten the steep rocks, this fern often grows in large sheets, with fronds of twice their usual length or more.

Less commonly there is a related species, the Tunbridge filmy fern, which forms slightly paler green and distinctly flatter patches, in the same kinds of habitat. The Tunbridge filmy fern is more sensitive to frost than its relative, and so in Scotland is found mainly below 500 ft. and in places within a few miles of the sea, reaching a northern limit in the south of Skye. By contrast, Wilson's filmy fern is widespread all over the western mountains, reaching at least 2500 ft. and extending to Shetland.

Often in the same rocky woods and glens as the Tunbridge fern is another fern of similar distribution but much larger size, the hay-scented buckler fern, best recognised by its rather light green crinkly fronds. It luxuriates in many places in this south-western district, and is another good indicator of a mild and humid climate. The soft shield fern, distinguished from the prickly kind by its softer texture, less glossy and more finely divided fronds, is a rarer plant, and confined to base-rich soils.

On the Isle of Arran, where most of these ferns flourish, there was once a much greater fern treasure in the beautiful Killarney Fern, the largest by far of the three British representatives of the great forest-dwelling tropical and sub-tropical family of filmy ferns. This plant hides its lace-like, dark green fronds in the deep shade of moist rock crannies and caves, but is an extreme rarity in the British Isles outside south-west Ireland. Its habitats abound in hill country and rocky coasts, but only very few and widely scattered patches have ever been found beyond the district after which it is named. In 1863, a postman had the good fortune to chance upon a colony in Arran, and two other patches were found elsewhere on the island later. But the news soon spread, and the island was scoured by fern enthusiasts from far and wide. The upshot was that the Killarney Fern has not been reported as seen in Scotland during this century. The postman's locality was certainly collected out, as he bitterly lamented, by a visitor to whom he had shown it, and who then had the audacity to publish the record as his own find!

What a rapacious bunch they were, these Victorian fern-hunters, with their pressed albums, wall frames, Wardian cases and conservatories. It is astonishing that some of the other rare species survived, so heavily were they plundered. On Arran, the royal fern was systematically depleted by the resident tinkers for sale to tourists, and was eventually eradicated except on inaccessible cliff ledges. Amongst the more common ferns, any difference in form suggestive of a variety was uprooted and carted off. Whilst ferns produce tiny wind-borne spores and are capable in theory of spreading over long distances, the rarer kinds have an extremely limited capacity for extending their existing ranges, and any reductions in their populations are more likely to be permanent.

But let us look now at the still lowlier and more ancient relatives of the ferns, the mosses and liverworts. These abound and luxuriate in the rocky woods and glens, and the specialist in these plants can find choice hunting grounds here. There are many localities where, without going more than a few hundred yards from the road, he can happily spend a whole day, poring over the rocks, banks and tree trunks.

Fig. 21 A patchwork of rock-encrusting lichens.
Extreme right
Fig. 22 Rose-root, a common cliff alpine, with rosebay willow herb on a ledge.

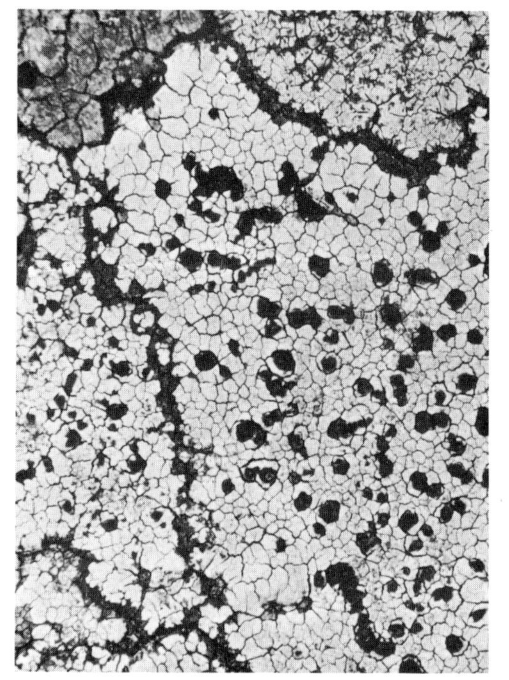

Fig. 23 A woodland fungus (Amanita citrina).

The mosses are of two growth forms; those with tightly packed upright stems forming dense cushions (*Fig. 10*), and those with flattened, creeping and usually much branched stems. Liverworts (*Fig. 11*) also are of two main types, the leafy kind, which look rather like some of the mosses, and the flattened, strap-like and fleshy forms.

This is not the place to go into the recognition of different species. For those who are interested, there are several books with useful introductions to the subject. Yet it is worth noticing the marvellous variety of colour, texture and form amongst the mosses and liverworts.

In places the rock surfaces and lower parts of the tree trunks are carpeted with these plants, in innumerable shades of green, gold, brown and red. Some are on the exposed surfaces, others in deep rock crevices; quite a number grow where water drips or splashes continually, or even submerged on stream rocks. Some belong especially to fallen and rotting logs. They variously grow as dense cushions and tufts; flat patches, wefts and flakes; tiny upright shoots or prostrate, creeping stems. Some are species confined to these western districts of mild and humid climate, and as with the filmy ferns, there are representatives of a much larger flora belonging to the humid tropics. A few are very rare, both in Britain and the world as a whole.

This is an important district for yet another group of primitive plants, the lichens, which have evolved through the combination in the same tissue of an alga and a fungus. There is a very large number of different kinds, varying greatly in appearance. Some simply encrust or etch themselves into the surface they cover, (*Fig. 21*) but others show considerable elaboration of the plant body, from the intricately-branched, bush-like form of the 'reindeer moss' *Cladonias* to the large festooning paper-like tatters of the tree lungwort.

Virtually all exposed rock surfaces become grown with lichens of the encrusting or etching type. The only exceptions are those very recently exposed, and rocks kept more or less constantly wet by sea or fresh water.

Some kinds are slow-growing. Interesting information about the actual rates of colonisation and growth has come from the examination of lichens on dated gravestones. The majority are extremely sensitive to atmospheric pollution especially by sulphur dioxide, and the richness of the lichen flora decreases rapidly with close approach to large cities and industrial centres. Conversely, western districts of Scotland, where the prevailing winds from the Atlantic maintain an atmosphere almost free from pollution, are extremely productive for these plants.

The trunks and larger branches of trees are an important lichen habitat. The most conspicuous are the tree lungwort and its relatives, forming flaky and scaly patches of green, olive and brown, especially on the older trees. Many of these large leafy lichens also grow on rocks. The matted wefts of old man's beard hang from trees, especially birch, in damp situations, and many other lichens grow mixed with the mosses and liverworts on rocks in woods and ravines.

The lichens have additional interest as a source of natural, vegetable dyes, derived from the acids which they contain. In byegone days they were widely used in the Highlands for dyeing tweeds and other woollens, and the species used went under the general name of 'crottles'.

The colours they yielded were mostly in the range from yellow to brown, but purples and reds were also obtained. Although yellow lichens tend to give yellow dyes, the colour is not necessarily related to that of the living plant. Some of the paler, crust-like kinds produce the most vivid tints, whereas the largest species tend to give more sombre browns. The dye colour varies also according to methods of extraction, and the time of collection, with August regarded as the best month for giving intensity of colour.

Finally, the woodlands are good habitat for the fungi themselves. Apart from the bracket fungi associated especially with birch, these plants produce their rather short-lived fruiting bodies—the toadstools—mainly in the autumn (*Fig. 23*). They are again plants whose identification requires patient study, with the aid of one of the several useful popular books on mushrooms and toadstools. Photographers will find them delightful subjects, with endless shades of colour covering almost the whole of the spectrum.

The woodland flora has the capacity to produce an ever-changing show of colour as the year moves round. The shiny yellow celandines are the first flowers to enliven the woodland scene in spring, followed by the white of wood anemones and wild garlic blended with a rich greenness of expanding young foliage. Bluebells and primroses (*Fig. 31*) are next, and the uncurling croziers of fern fronds have unusual elegance. Borrer's male fern, with its thick felt of chaffy, red-brown scales is an especially attractive plant (*Fig. 29*).

Each month brings new colours until autumn provides a climax, ending with leaf-fall. Even in the drabness of winter, the mosses and liverworts stand out intensely green and lush, and the bare trees have themselves a certain richness of hue.

Where the coast bounding the sheltered inlet of Loch Fyne swings round suddenly to face the great surge of the open Atlantic at the Mull of Kintyre, there is a change of scene.

The ground by the sea steepens into high cliffs, and woodland becomes confined to narrow glens running back into the hills, apart from the patchy growths of scrub on the less vertical headlands. Heathery and peaty moorland extends right to the top of the cliffs, and the impression is of hill country running down to the sea. The coastal rocks and banks have an abundance of thrift, sea campion, sea plantain, bucks horn plantain, sea spleenwort, scentless mayweed, English stonecrop and grasses. The short heaths and grassland on the cliff-tops have local abundance of the spring squill, an attractive relative of the bluebell. On less precipitous slopes and ledges, above the spray zone, there are woodland communities of herbs, with red campion, bluebell, greater stitchwort, great woodrush, meadowsweet, angelica (*Fig. 24*), hogweed, sorrel, valerian, and, much more sparingly, the beautiful wood vetch (*Fig. 13*). Brambles and the more common ferns are plentiful, too.

This is one of the southernmost localities for the lovage, a maritime plant almost confined to Scotland. The northern character of these coastal headlands is most interestingly revealed by the presence of a distinctly upland element in the flora.

The fleshy, yellow-flowered roseroot (*Fig. 22*), so familiar on mountain cliffs, grows in big bunches on the rocks in various places, whilst to the north of the Mull point is an outcrop of lime-bearing strata with a remarkable collection of alpines. The mountain avens is here in some abundance, with purple, mossy and yellow saxifrages, and a colony of an oxytropis, or mountain milk vetch, which seems to be neither the yellow nor the purple species but a form half-way between the two. These alpines have survived at such low levels because the windblasting has prevented woodland from smothering the cliff faces and summits, and the low summer temperatures maintained by the influence of the sea are favourable to the growth of such plants.

Above the sea cliffs of Kintyre the moorlands rise for the most part smoothly to a central spine. There is great abundance of the common upland plants—heather, bracken, cotton grasses, deer sedge, purple moor grass, mat grass, heath rush and so on—but rather little in the way of special interest. The mountains of the Clyde area and the district of Lorne to the north are, in general, rather limited in the variety of their Arctic-alpine flora. The granite hills of Arran are rugged and spectacular, but distinctly lacking in botanical appeal, with a few plants of poorer soils such as alpine ladies' mantle, least willow, mountain sedge and spiked woodrush as their most distinctive species.

The still higher mountains above Loch Etive and Loch Awe are rather similar in their lack of floral variety, but one limited limestone outcrop hereabouts has a fine species in the purple oxytropis, one of the most beautiful of all plants growing wild in this country (*Fig. 12*), and the rarer, less conspicuous Arctic sandwort.

Serpentine outcrops above Glendaruel have mountain avens, purple and yellow saxifrages, and some of the more common alpines, but the hills of Cowal tend to be grassy sheep walks and are not especially productive for alpines.

The mountains of the south-west Highlands lying farthest inland, on the borders of Argyll and Perthshire, are those with the greatest botanical interest and a detailed look at the montane flora of the district can most conveniently be made here. These hills have most of the plants to be found in the coastal uplands, but a great many more besides.

Before leaving the coast of the south-west, we should be aware of a further group of plants which are rare or very local in the Highlands and important on that account. The chances of finding them casually are small, but all of them could well turn up in new places in the district, as a reward for careful search.

The shingly shores of Arran provide one of the southernmost British localities now known for the handsome oyster plant (*Fig. 58*), which has retreated northwards by loss of more southern colonies during this century. Arran also has the Isle of Man cabbage, a rather ordinary-looking yellow flowered crucifer which is not known outside Britain and is extremely local even there. Some warmth-loving plants which belong especially to the Mediterranean countries reach their northern limits hereabouts—the lanceolate spleenwort fern on rocks on the east coast of Kintyre is such an example.

The rustyback fern is not quite so rare. As elsewhere in this country, it grows mainly from the crevices of mortared walls, in company with the much more common maidenhair spleenwort and wall rue. The delicate maidenhair fern has never been found in Scotland, but it could well occur in this district, on seacliffs formed of lime-bearing rock, for it is known from such places on the Antrim coast only a short distance away across the Irish Sea. The great horsetail and its frequent companion, the striking pendulous sedge are other southern plants reaching this district. Two attractive herbs which belong especially to Wales and south-west England also have a foothold: the tiny ivy-leaved bell-flower in marshy, grassy places, and the curious wall pennywort or navelwort on dry, acidic rocks and walls.

Many people enter the Highland Region by crossing from Perth to Argyll just past Tyndrum, heading either for Fort William or Oban. This is exciting terrain for the botanist. To the east rolls the long Perthshire Breadalbane range, culminating in Ben Lawers (3984 ft.), the most famous British hill for alpine flowers and now owned by the National Trust for Scotland. Several hills at the western, Argyll end of this great range are almost equally rich botanically.

The secret of this richness lies in the thick beds of calcareous mica-schist and limestone in the Dalradian rocks which form these mountains. The schist is soft and weathers readily to give fertile soils full of silvery flakes. It is exposed extensively in cliffs on Beinn Laoigh (3708 ft.), the fine peak appearing westwards of Crianlarich, and in the lofty hills east of the railway beyond Tyndrum—Beinn Dorain, Beinn an Dothaidh, Beinn Achaladair and Beinn a'Chreachain. Certain alpines are remarkably constant in their appearance on each of the main hills, and the differences between one mountain and another are more in the varying abundance of certain plants.

At first sight, these hills give little hint of the treasures they hold. Bleak and bare, with gully-seamed slopes in places, they have been badly knocked about by man and his animals down the centuries.

Surviving birches and rowans cling in open growth to the lower cliffs. There are stands of old Scots pine at Cononish, Loch Tulla and Crannach amongst these hills, but this is an otherwise deforested country. Sheep and moor fires have made a long onslaught on the lower slopes, with the result that plants such as deer sedge and purple moor grass are in great quantity.

Heather is common but generally poor in growth, except in some of the rockier places where it is tall and bushy. The rainfall is heavy and the gentler slopes and flats at the foot of the hills are boggy. The bog myrtle, a low shrub which grows much taller when

Fig. 24 Angelica, one of the larger herbs flourishing in ungrazed places, here on slopes above the sea.
Extreme right
Fig. 25 Butterwort, an insectivorous plant of moist ground.

Fig. 26 Starry saxifrage, growing in cushions of apple moss on a rock ledge.

neither burned nor grazed, is common here. Crush a leaf or two—they have an aromatic fragrance to associate thereafter with these northern boglands. Cross-leaved heath is another abundant and even smaller shrub of the moist ground. Bog-mosses (*Sphagnum*) abound, and rooted in their carpets are plants such as bog asphodel, round-leaved sundew, marsh violet and common butterwort. Drier ground has grasses, especially bent, fescue, mat-grass, sparse heather and bilberry, and a group of herbs with small but colourful flowers; the yellow of tormentil, blue of milkwort, pink-red of lousewort and white of heath bedstraw. Devil's bit scabious is common, but flowers best in ungrazed places. Dips and hollows where drainage water collects have growths of common rush, sharp-flowered rush, star sedge, common sedge and common cotton-grass. The heath rush, with its dense and wiry 'bird's-nest' rosettes flourishes on peaty soils.

The approaches to the high ground thus tend to be over terrain which is rather limited in botanical variety, and all the plants so far named can be seen well to the south of the Highands, even on the damper heaths and bogs of southern England. This is only the beginning, though, for the bases of these hills lie at a low elevation (about 500 ft.), and the alpines belong especially to the ground beyond the natural tree limit, which here at least is a thousand feet higher.

The bigger cliffs are the places to aim for, but with all due caution, for much of the rock is frighteningly rotten and unreliable for climbing. Not all the crags are formed of the productive calcareous schist, but the plant seeker will have no doubt when he has struck the right kind of rock; almost any sizeable outcrop of the Lawers schist above 1500 ft. will prove to be a veritable natural alpine rock-garden, and the main precipices are quite spectacular in their wealth of plant life.

The mountain plants are not all cliff dwellers: some have a wide range of habitat. A few appear low on the slopes where water draining from the rich rocks influences the ground below, giving damp grassland, little sedgy marshes and flushes with an open muddy or gravelly surface. These places have the curious viviparous bistort, whose flowering stems have a terminal spike of white florets, passing below to red bulbils which fall off and grow directly into a new plant. The tiny alpine meadow-rue is often abundant, and the open flushes are a typical haunt of the yellow saxifrage and the rather less common purple saxifrage. Climbers active amongst these hills in March and April will have the delight of seeing the purple saxifrage in bloom, for it is the earliest of the mountain flowers, and its marvellously rich pink-red blossoms (*Fig. 46*) are a great adornment to the bare mountain cliffs when the other vegetation is showing little sign of growth. This hardiest of plants forms its buds beneath the snow and ice of winter, as befits one of the select few Arctic species which, growing at 83° north, are the most northerly plants in the world.

During really wet weather the little streams and rills rise to flood their banks and water courses over ground normally quite dry. Such places have abundant flat patches of the alpine-ladies' mantle (*Fig. 19*), with its deeply lobed leaves, silvery-silky underneath, and clusters of tiny yellow flowers. With it there are usually dense mats of wild thyme, with flowers of deep pink and leaves intensely fragrant under one's tread.

The long steep slopes which are the usual route to the cliffs have mostly a cover of grassland or bilberry, providing a remarkable contrast with the vegetation of the ungrazed ledges above. These are fascinating places for the plant hunter.

The cliffs of the mica-schist mountains are typically broken up, with tier after tier of rock separated by banks and ledges of varying steepness and accessibility. There is a prevailing lushness about the vegetation on these more broken escarpments. It seems almost to cascade down the precipice, in real hanging gardens. The steep rock faces have all manner of niches, cracks, corners, pockets, scoops, ledges, overhangs and caves in which plants can find anchorage.

Certain of the alpines are confined to rocks where the combination of steepness and gravity forever resist the build-up of soil and the entry of competitors. Some of them sprout, miraculously and in full vigour, from the tiniest crevices of sheer rock. Though smooth faces may bear growths only of moss, liverwort, and lichen, it is remarkable how luxuriant a cover of vegetation can develop on near vertical walls. Certain cliffs have some incredible curtains draping the sheer rock, with masses of yellow, mossy and purple saxifrages, alpine ladies' mantle, alpine scurvy grass, moss campion, mossy cyphel, viviparous bistort, alpine meadow rue, alpine mouse-ear, northern yellow-rattle, three-flowered rush, black sedge, hair sedge, alpine meadow grass and viviparous fescue. These are the true alpines, but mixed with them are familiar lowland plants such as early purple orchid, frog orchid, common twayblade, grass of Parnassus, dandelion, harebell, purging flax, heath violet, thyme, eyebright, meadow buttercup, moonwort, kidney vetch, wood anemone, hairy rock-cress, broad-leaved willow herb, wild strawberry, herb robert, ribwort plantain, flea sedge, yellow sedge, carnation grass, red fescue and sheep's fescue.

Some of the more robust alpines can find a foothold on steep, bare rocks, but they flourish best and reach their most vigorous development on the broader ledges and pockets where a deep layer of soil has built up (*Fig. 27*).

They include the rose-root, mountain sorrel, alpine saw-wort, northern bedstraw, holly fern, and stone bramble. Typically in their company is a group of plants which may be seen widely in Scotland on certain river banks and roadside verges, in hay meadows and some hill woods; such as the globe flower (*Fig. 17*), wood cranesbill, melancholy thistle, marsh hawksbeard, water avens, meadow sweet, dog's mercury, angelica, hogweed, common ladies' mantle, valerian, red campion, golden rod, devil's bit scabious, wood vetch, great woodrush and tufted hair grass. This dense herbaceous growth makes a particularly striking contrast with the sheep and deer-grazed foot of the cliffs, where tufted hair grass, great woodrush and bilberry are the plants most in evidence, though careful search will usually reveal grazed-down and stunted remnants of the ledge species.

The exclusion of the animals by a fence produces a remarkable change quite quickly, as may be seen below the cliffs of Creag an Lochain to the west of Ben Lawers, where the Nature Conservancy Council have established experimental 'exclosures'. The grazed down herbs have grown tall and flower well, and they are beginning to re-assert themselves against the supremacy of the grasses. These medium to tall herb communities were once extremely widespread on the richer soils, and they originally represented the herbaceous component of mountain birch-hazel woods and willow scrub. Most of these plants may be seen in this role in Scandinavia, where so much of this sub-alpine woodland and scrub still remains.

The alpines mentioned so far may be found on nearly all the main cliff ranges of these Argyll and Perth Dalradian schist mountains, and often in great quantity. Some others are less constant in appearance, occurring plentifully in some localities, but sparsely or not at all in others. The mountain avens is one of them, a beautiful creamy-flowered dwarf shrub with glossy, dark-green scalloped leaves and knotted, trailing woody stems. It is another of the widespread plants of the great Alpine ranges and the Arctic barren grounds, right around the world. With it frequently is the net-leaved willow, recognisable by the deeply impressed veins in the shining, rounded leaves (*Fig. 32*). The mountain willow is especially abundant on some of these hills, but it hardly occurs outside the Breadalbane and Clova districts in Britain. Whortle and downy willows are somewhat local, and the rarest of these mountain shrubs, the woolly willow occurs in this district only in a very few places within the Perthshire portion. It is a strikingly beautiful plant (*Fig. 39*), with silvery down on its leaves, and male catkins which are large, silky and golden.

Among the local herbaceous alpines are the hoary whitlow grass and its rarer relative, the rock whitlow grass (which are not grasses at all, but small, white flowered crucifers), alpine cinquefoil, alpine pearlwort, alpine bartsia, northern rock-cress, rock speedwell, alpine saxifrage, rock sedge, glaucous meadow grass

(*Fig. 20*), meadow oat grass and blue moor grass.

Two rare and attractive ferns grow in scattered localities and contrasting habitats. The delicate mountain bladder fern, with distinctive broadly triangular fronds, belongs to the damp, shady ledges, banks and crannies, where it is often with the moisture-loving yellow saxifrage. Alpine woodsia (*Fig. 30*), sprouts its dainty little tufts from quite dry, steep and often bare rocks. Both these ferns have suffered from collecting, the second particularly so. Some of its colonies have been depleted even within the last ten years by irresponsible fern collectors.

Several other alpines—the snow gentian, for example, and alpine fleabane, alpine forget-me-not, drooping saxifrage, snow pearlwort and mountain sandwort—are to be found mainly or only on Ben Lawers and some of its closest neighbours in Perthshire. Some of these are amongst the rarest plants in Britain though, as with nearly all our alpines, they may be seen growing plentifully on the great mountain ranges of continental Europe. Even our best mountains can hardly be compared in botanical richness with those of the Alps, Pyrenees, Jura and Norway but the Highlands offer a great deal of variety within a small compass. Some of the particular combinations in which plants occur here are not to be found anywhere else in the world.

These western Breadalbane mountains are tremendously rich in mosses and liverworts, and one of the most beautiful is *Orthothecium rufescens*, which forms large, silky and deep-red tufts and cushions on the high rocks (*Fig. 16*). The ground below the cliffs is in many places carpeted by dwarf alpine herbs, of which the flat green patches of moss campion and mossy cyphel are especially prominent, grown and mixed with many of the cliff-face species. Wet places on the high slopes often have an abundance of the russet sedge, and open flushes have yellow and purple saxifrages, Scottish asphodel, three-flowered rush and, more rarely, two-flowered rush, chestnut rush and Kobresia, a small sedge-like plant.

Snow lies late on many of these upper slopes. There are many examples of distinctive late snow-patch vegetation, with first a prevalence of mat-grass and tufted hair grass. This passes, with increasing duration of snow cover, to carpets of moss and small liverwort forming a velvety felt over the soil surface, grown with small alpines such as alpine ladies' mantle, Sibbaldia, least cudweed, alpine willow-herb and spiked woodrush.

On high slopes less strongly influenced by long snow cover there is a mixture of short bilberry, crowberry, mat grass, bent and alpine ladies' mantle, with increasing amounts of the dense growth of woolly fringe moss. As the slopes flatten out into the mountain summits, the fringe moss takes over almost completely, forming thick carpets with only a sparse growth of the other plants (*Fig. 70*), together with rigid sedge, least willow and three-pointed rush (*Fig. 53*). Here and there, the fringe moss heaths have flat patches of moss campion, whose little pink flowers enliven the drabness of the scene at mid-summer. Lichens, mosses and liverworts are much in evidence, especially on the rocks and in soily places. The general sparseness of the vegetation is in keeping with the bleakness of the climate on these high tops.

Beyond Bridge of Orchy the high hills of western Breadalbane and the Black Mount Forest give way to a vast moorland basin of completely different character. After the road climbs the brae above Loch Tulla with its pinewood fringe there opens out a spacious vista of loch and moor backed by distant, higher hills—the great Moor of Rannoch.

Though crossed by the main A82 road on the west side, and the west Highland railway along the eastern edge, this is a wild and desolate tract of country known to very few. It is a place to savour the wilderness experience, but do so with a map, compass and an eye to landmarks and weather conditions, for there is no great difficulty in becoming truly lost. In fine weather, the great mountains surrounding the Moor give direction, distance and scale to this expanse of over 60 square miles. When the clouds hang low on these hills, or the horizon becomes blotted out by driving rain, this can be an unfriendly, frightening place; yet it is such conditions

Fig. 27 The hanging gardens of the mountain cliffs – broad ungrazed ledges and banks with dense herbaceous growths, and open rock faces with small alpines.

Fig. 28 Foxglove – a common and handsome but poisonous plant of poor soils.
Fig. 29 Borrer's male fern – one of the most attractive ferns of both woodland and open ground, here unfolding its fronds in spring.

Fig. 30 Alpine woodsia – one of the rarest British ferns, reduced by collectors and confined to a few elevated mountain cliffs in the central Highlands. It is growing here with roseroot, purple and alpine saxifrages.

which seem the more appropriate. For this is a kind of wet desert, an inhospitable expanse of peat bog and open water which remains one of Nature's last refuges.

The moor is an undulating glacial plain: the deeper troughs and holes have filled with water, whilst deep layers of blanket bog have formed over the general floor. Only the numerous morainic knolls and rock hillocks have remained dry. Apart from the conifer plantations of the Forestry Commission which have begun to march across the western side this is a treeless area. The drier moraines carried pinewood until this was obliterated during the relatively recent period of forest clearance, and only a few islands in the larger lochs carry tiny remnants of wood and scrub to tell the story of human impact.

Much of the Moor is flat 'flow' bog, carpeted by *Sphagnum* and much too wet to have carried trees in recent times. Yet here and there, patches of bog in vulnerable places have become severely eroded. They nearly all reveal the remains of large pine trees (*Fig. 3*), sometimes lying at the bottom of the peat but in other places at a higher level. There are well preserved stumps with spreading roots, and fallen trunks in places, bleached and exposed now to final decay. The oldest have an age of around 7000 years, but some are much more recent.

The drying edges of these eroding bogs have large banks and mounds of the woolly fringe moss, one of the most characteristic plants of the western Highlands. Where the ground has been burned repeatedly there is a prevailing cover of deer sedge and ling heather, giving way on the dry moraines to ling and bell heather with the upright yellow-flowered petty whin and creeping flat patches of bear-berry. The sombre-looking patches of flat, wet bog prove on closer inspection to have quite a varied appearance The bog mosses which form the prevailing ground cover are of several different kinds and colours. Some, of an intense green, occupy wet hollows and pools, whilst those forming lawn-like surfaces range from pale green through yellow to deep wine red. In places the surface has a hummocky structure, and two of the less common bog-mosses form particularly large, domed cushions of an orange or brown hue. Rooted in the Sphagnum carpets is a rather open growth of other plants: sheathing cotton grass, deer sedge, purple moor grass, cross-leaved heath, the yellow flowered bog asphodel and the two sundews, round-leaved and English, with their sticky, insect-catching leaves.

The pools of open water, and the wet hollows have a particular group of plants. The handsome bog bean flourishes here (*Fig. 51*). and there is abundance of common cotton grass, distinguished from the other kind by its multiple instead of single heads of cotton–the fruiting stage of both. Dense patches of the pretty little bog sedge, with its delicate nodding spikes, are frequent, and there is some abundance of the tiny lesser bladderwort.

This area is the only one known in Britain now for a rare and rather inconspicuous rush, accordingly known as the Rannoch rush, which grows sparingly in these wet bog moss hollows. It is a plant widespread in the great bogs of northern and central Europe, and its distinctive remains have been found in peat deposits in many parts of the British Isles. Its decline and eventual restriction to the area of Rannoch Moor is one of the mysteries of plant distribution in this country. Its habitat is still widespread enough, but it may have a need for some subtle special conditions, perhaps of climate, which only the most careful research could reveal.

Here and there in the Sphagnum carpets are patches of the dwarf birch, though this little shrub also grows on the shallow, drier peat at the edge of some of the moraines. It is another plant once much more widespread, and Rannoch is one of its most southerly locations in Britain. Again, habitat is widespread, and whilst the dwarf birch has clearly suffered loss of ground through moor burning and the associated drying and erosion of blanket bogs, this is not the whole story, and it again seems that obscure climatic factors may be involved.

The wet bog moss areas have a water table virtually at the ground surface, and the bog-trotting naturalist needs rubber boots to remain dry-shod for long. These are not

Fig. 31 Primrose – the familiar wayside and woodland flower which grows profusely in many parts of the Highlands.

dangerous bogs, though, for all their forbidding appearance, and one soon learns to avoid the step which will result in a water-filled gumboot. Sheep sometimes become stuck or drown in pools, but probably usually through falling through thin ice or snow during winter, and this could be a hazard for the unwary traveller, especially in the dark. The unfortunate sheep which have expired so miserably leave their mark after the last traces of wool and bones have disappeared—there is an unusual lushness and greenness about the vegetation in the pool, for the plants have benefited from the unaccustomed boost in fertility, as the nutrient cycle has turned full swing. For these are habitats of low fertility, where the essential elements—calcium, potassium, nitrogen and phosphorus—are at a premium, and the growth even of bog plants is limited by the shortness of their supply. On these deep peat bogs, the living surface is completely separated from the underlying mineral soil, and depends largely on the small amounts of nutrients which fall on it as dust or in rain water.

In places, slight slopes and hollows cause a movement of the water through the surface layers of peat, and these seepage tracks are marked by an abundance of sedges, notably bottle sedge, slender sedge, star sedge, and bog sedge.

Where there is open water or bare peat sludge the intermediate, greater, lesser and bladderworts are found, and other typical plants of these seepage areas are lesser spearwort and bog pondweed. These flushes and runnels often lead sooner or later into lochans or bigger lochs, and these in turn feed bigger outflowing streams and rivers.

Some of the smaller lochans are, however, in completely enclosed basins, probably representing ancient 'kettle-holes' where large chunks of ice were stranded amidst the developing moraines as the glaciers dwindled, leaving deep holes when they finally melted away. Floating rafts of *Sphagnum* have developed around the edges, in places, and here the delicate stems of cranberry creep over the moss. This is the habitat of the rare tall bog sedge, and the more common white beak sedge. Some lochans are fringed by bottle sedge swamp, and the open water plants include bogbean, water milfoil, white and yellow water lillies (*Fig. 69*). Stony edged lochs have abundance of shoreweed, and in the shallows, the water lobelia shows its pretty little pale blue flowers above the surface (*Fig. 14*). Quillwort is another of the shallow lake edge plants with clusters of spiky leaves, and the water awlwort is a rarer member of this group.

The islands in the larger lochs—Loch Laidon and, on either side of the main road, Loch Baa and Lochan na h'Achlaise—tell us something of man's influence on these uplands. Notice their long, shaggy heather and clumps of pines and birches, contrasting with the short vegetation of the treeless shores and surrounding moorland. How different a scene this must have been two thousand years ago, before the fire raisers and their animals set to work in earnest. Not all loch islands survive unscathed, for deer can swim to most of them, and the pyromaniacs have seldom missed a chance to make an offering to the fire-god.

The hills overlooking Rannoch Moor to the west are mainly of granite. Though they have rather more heather and other dwarf shrubs, they are botanically rather dull by comparison with the schistose ranges farther south. On the Black Mount tops such as Clach Leathad there is abundance of plants such as dwarf azalea, bearberry, three-pointed rush, spiked woodrush, least willow and least cudweed, and at lower levels in Coire Ba is great abundance of the interrupted club-moss.

The great andesite bastion of the Buachaille Etive Mor standing sentinel at the head of Glen Etive is tremendously impressive in its rock scenery, but this is a climber's mountain and there is even less here for the botanist. Farther on, where the mountains begin to enclose the road, and the last bogs of the Moor are left behind, there is richer ground, but it has to be sought carefully.

Glencoe is the most spectacular mountain scene to be viewed from any road in the British Isles. It is a classic U-shaped glaciated valley, with the triple precipitous truncated spurs of the Three Sisters on Bidean nam Bian facing the long, gully-riven escarpment of Aonach

Eagach to the north. This again is climber's ground, but besides the hard andesite walls, there are outcrops of calcareous rock where the Breadalbane flora appears again on a limited scale.

Some of the more common mountain plants occur here at low levels, on the first rocks above Loch Achtriochtan—roseroot, mountain sorrel, yellow saxifrage, alpine ladies' mantle, interrupted club moss, viviparous bistort and alpine meadow rue. The higher corries have many of the lime-loving alpines of the Bridge of Orchy hills, including rarities such as mountain bladder fern. There is a good deal of two plants uncommon on the Breadalbane range, the Arctic mouse-ear and alpine tufted hair grass, and bleak bare rocks at high elevations have two rare Ben Lawers plants, the drooping and Highland saxifrages.

The next hill to Bidean, the much lower rounded lump of Meall Mor shows marvellously well the influence of geology on plant life in the Highlands. Its Glen Coe face shows a series of low, broken cliffs which doubtless hardly receive a second glance from the rock gymnasts heading for the great walls around Ossian's Cave higher up the glen. But it is formed of limestone and the botanist will soon realise that this is a hill of some quality. Mountain avens is in abundance, often growing down into grassland, and there is much whortle willow, as well as many of the more common alpines mentioned for the Breadalbane hills. There is also the lesser meadow rue, a rather uncommon plant of the Highland mountain rock ledges.

Beyond this, the glen opens out, and it is only a short run to Glen Coe village, and Loch Leven, a narrow west-coast arm of the sea. We are back in wooded country again, and amongst plants which belong to the mild, western seaboard. Only a few miles from the alpines in the late snow hollows high above Glencoe, there is the Tunbridge filmy fern flourishing in rocky woods near Ballachulish. There are interesting woods along the coast from here south to Oban and Knapdale, and limestone near Appin has the beautiful sword-leaved helleborine. *(Fig. 61)*

Fig. 32 Net-leaved willow, one of the rarer alpines of the mica-schist mountains, growing with moss campion, purple saxifrage and viviparous bistort.

The Central Highlands

The other usual approach to the Highlands is through east Perthshire, via the A9 road or by rail on the route of the old Highland Railway from Perth to Inverness. Beyond Pitlochry and Blair Atholl the dense woods of the lower Garry valley thin out and there is a broad sweep of smooth, rounded hills, not unlike many parts of the Southern Uplands in general form, but revealing their deceptively greater height and colder climate by the numerous snow patches lingering on the upper slopes into early summer. These heathery mountains close in around Dalnaspidal, and just beyond, the gradually ascending road reaches its highest point of 1500 ft. in the Pass of Drumochter, and there enters the Highlands and Islands on the Perth–Inverness march.

This is now a bleak, almost treeless, scene of elevated grouse moor and deer forest, with steep slopes of heather rising another 1000–1500 ft. above the road before levelling out into high spurs and plateaux on either side. In the trough of Drumochter are tracts of bogland interspersed with hummocky clusters of dry heathy moraines. This is a good place for an introduction to the mountain flora, and a short wander from one of the various parking places beside the main road can be instructive.

The sides of little streams and rills are especially productive, for there are richer soils here than on the dry hill slopes. In the grassier places, right down to the road edge, there is abundance of alpine ladies' mantle and, on bare, earthy banks, beside both road and railway in this, the Badenoch district, there are characteristic flat, trailing green carpets of the bearberry, a northern, evergreen dwarf shrub. It is distinguished from the still more widespread and plentiful cowberry by its prostrate instead of upright stems, and more tongue-shaped instead of oval leaves, which also have more conspicuous networks of veins on their surface. Both these relatives of the paler, deciduous-leaved bilberry have whitish bell-like flowers, and bright red berries (***Fig.** 37*).

Bilberry forms conspicuous green patches in slight hollows where the winter snow lies longer than on average. Here it is sometimes accompanied by the rarer bog whortleberry, which has blue-green leaves, and rounded instead of angled stems. Both plants produce blue-dark purple berries of a highly edible kind, but those of the rarer species are found rather sparingly in this country, which is perhaps as well, in view of their allegedly intoxicating properties.

Mixed with the bilberry are other plants—the pretty little white-flowered dwarf cornel (*Fig. 34*), whose petals are really sepals; the hard fern so familiar in the rocky western woods; and, more locally, the trailing interrupted clubmoss, the most alpine of this ancient group of fern-allies. On rockier ground, especially, bilberry is accompanied or replaced by crowberry, which here is usually the more robust, greener and more trailing northern form, with hermaphrodite flowers, quite distinct from the redder stemmed common crowberry of more southern hills, carrying male and female flowers on different plants. To distinguish them most readily, search for the clusters of little, shiny blackish berries in summer, and pick one carefully, If there are three stamens clinging to its base, you have the hermaphrodite kind; if not, the other, more widespread form.

Where snow lies still longer into the spring,

bilberry gives way to patches of the pale mat grass, though this plant occurs in great quantity on southern, sheep-grazed hills where quite different conditions determine its abundance. It is in a mixed community on one of these Drumochter hills, where bilberry grades into mat grass, that the very rare little shrub known as Menzies' heath (or blue heath) has long been known to grow. Until quite recently, it was supposed that this was the only British locality for the plant, but Ron MacBeath, searching remote hills in this district, has found it sparingly in two other places.

The habitat is so widespread that the extreme localisation of the plant is something of a mystery, and gives the feeling that it could yet be found in still other new spots. When in flower, its large pink-purple bells are conspicuous, but at other times it is easily overlooked, for the foliage is passably like that of the crowberry.

The damper heather slopes on the shadier aspects have two characteristic plants. One is a tiny and inconspicuous orchid, the lesser twayblade, which grows rooted in the carpets of moss under or between the lankier growths of heather; the other is the cloudberry (*Fig. 44*), a relative of the bramble, but with soft, dark green leaves, lacking thorns and trailing stems. The white flowers, which soon lose their petals through wind and rain, have male and female separated on different plants. The fruit has the form of a blackberry, and passes through a red stage and then to golden-yellow in the fully ripe berries.

This is one of the important fruits of the Scandinavian countries, for it is a true Arctic plant and grows there in great quantities. Although a common plant on our northern moorlands too, it seldom produces fruit in real abundance with us. The berry has a slightly insipid taste, but with a distinctive tang entirely its own, and some people find it attractive. A shepherd with whom I stayed once, invited me to try some cloudberry jam, with the comment, 'It has a wild taste, and I like it'. I had to agree with him.

The dwarf birch is a rather rare plant of the bogs in this area. These plants all belong to acid soils, and some of the older botanists called them 'peat alpines'. The Drumochter hills are formed mainly of poor rocks, but in a few places there are outcrops containing lime. Close to the road there is a little stream with shingle edges in which grow purple and yellow saxifrages, and mountain sorrel. Higher into the hills are crags with greater quantity of these plants, together with mountain avens, moss campion, holly fern, downy willow, whortle willow, black sedge, rock sedge, hair sedge, alpine saw-wort, mossy saxifrage, alpine mouse-ear and hoary whitlow grass. The cliffs soon give out, though, and this is not the best area for seeing alpines with a fondness for lime-rich soils.

On the upper slopes where snow lies into late spring there are other alpines such as least cudweed and Sibbaldia. Little marshes and diffuse mossy springs have a few rare plants such as mountain bog sedge and alpine foxtail grass. Blanket bogs with abundant cloudberry straddle some of the high watersheds, and the view eastwards across the great Forest of Atholl is of a vast dissected plateau-land, with great expanses of this kind of vegetation, variously gullied and eroded in many places. Dry, windswept shoulders and spurs leading to the tops show gradual decrease in the stature of heather, until finally this occurs as a dense, flattened mat pressed tightly to the ground, grown with lichens and accompanied by dwarf azalea, crowberry, and least willow. In places there are alternating strips of dwarf heather and bare gravel or sand running along the contours.

The heather eventually gives out, and the broad, flat tops of these hills are mostly covered by carpets of the woolly fringe moss. This is a remarkable plant, for it grows in a wide range of different habitats, from sea level to the highest tops, and covers altogether a great many square miles of Scotland. It is an almost cosmopolitan species but thrives especially in the more humid parts of the world. Golden-green when moist and grey when dry, it grows in sheets over these summits, deep and yielding to the tread, smooth in some places but forming regular hummocks in others, with some resemblance to an undulating sea. The hummocks, which are usually about 3–4 feet across and up to 1 foot or so high, appear to have developed through frost-thaw movements producing a sorting of the surface soil and stones. Where the ground begins to slope, the hummocks run together to form ridges. These features correspond to the stone-nets and stripes so characteristic of the high Arctic regions, and found also in places on the bare soil and rock debris of the higher British mountains.

The fringe moss heaths have an abundance of mountain sedge, with sparse, dwarfed growths of bilberry, cowberry and a few grasses. Least willow is plentiful where the moss carpet is thinner or partly eroded. The flora of this ground is very limited, but it is interesting terrain to walk over—a somewhat eerie wilderness, deserted and silent, except for the occasional weird croaking of the ptarmigan.

On windswept high ground the fringe moss is nature's compass, useful when mist descends, for its flattened stems grow in parallel rows all pointing directly away from the prevailing wind; that is their tips all point to the east-north-east. Check it against a real compass if you wish to be convinced. The same is true of other plants which grow with flattened stems on the high tops, such as dwarfed heather and crowberry.

From Drumochter there drains eastwards one of the headstreams of the River Spey, and as the strath opens out past Dalwhinnie, trees begin to appear again in increasing number, though they are mostly birch. The green patches of bearberry on the roadside banks here are conspicuous, and a disused quarry floor has a remarkable quantity of yellow saxifrage, mixed with wild thyme in a delightful pattern of yellow and deep pink.

Behind Newtonmore rises the great tableland of the Monadhliath, the Grey Mountains, similar in topographic and botanical character to the hills of Atholl and Drumochter, but still more desolate, inaccessible and untrodden. Their glens have some deep, rocky ravines and there are a few high corries, but much of the high interior is covered by weary eroding expanses of blanket bog and the flora is rather limited. There is a good deal of dwarf birch locally, and southern occurrences of the alpine bearberry, whilst a remarkably fine colony of holly fern in one place suggests that the area was not much worked by the Victorian collectors. The lower slopes have patchy birchwoods, but these are more extensive lower down Strathspey, in the area around Aviemore.

In its middle reaches, the Spey becomes a sluggish river, meandering through a broad alluvial valley, in marked contrast to the swift, rocky courses of its headstreams. Below Kingussie it has been embanked artificially, and is flanked on both sides by a great tract of swampland, the Insh Marshes, over which the Royal Society for the Protection of Birds have established a nature reserve.

The Insh Marshes are the biggest expanse of fen in Scotland and are exceeded in size in Great Britain only by the swamps surrounding the Norfolk Broads. The waterlogged valley bottom here contains deep layers of peat built up by the swamp vegetation over thousands of years, but compared with the Norfolk Broads, the water is rather low in amounts of nutrient, and many plants of southern fen are absent. There are large areas of sedge, and other swamp plants such as marsh marigold, bogbean, marsh cinquefoil, marsh bedstraw, skullcap, lesser spearwort, marsh willow herb and horsetail. In places are dense, tall beds of reed and reedgrass, and at the drier edges the marsh vegetation grades into damp meadow grassland, with tufted hair grass, rushes, meadowsweet, valerian, devil's bit scabious and creeping buttercup. Below Insh village is bog and wet heath with a variety of bog-mosses, purple moor grass, bog myrtle, heather, cross-leaved heath, deer sedge and bog asphodel. Lower down, towards Loch Insh, there are thickets of willow which have colonised the swampland, and in places alderwood has also developed. Large open pools in the marshes have rafts of the beautiful white and yellow water lilies, bogbean, floating bur-reed and bulrush.

Near Kincraig the Spey broadens out into Loch Insh, which has shallow edges variously

Fig. 33 Twinflower, one of the notable plants of the north-eastern pinewoods.

Fig. 34 Dwarf cornel, an alpine of poor soils, growing amongst bilberry and crowberry.

grown with sedge, several kinds of pondweed, quillwort, water milfoil, shoreweed and bulbous rush. Still lower down, at Lynwilg, a sealed-off loop of the Spey under the hill of Torr Alvie has developed a superb swamp with sedge beds, and great sheets of bogbean, white and yellow water lilies. The northern water sedge is one of the abundant swamp plants of the Spey valley, along with the more common bottle sedge.

The sedges may seem a forbiddingly difficult group to become acquainted with, but they are not really so. Many species have quite distinctive habitats and distributions, and a little patience with a key to their identification will be rewarded. After that, it will be a challenge to see how many new kinds can be found. There are other lochs around Aviemore —Alvie, Pityoulish, Morlich and Loch an Eilean, all with a similar and rather limited range of aquatic plants, reflecting the lack of dissolved substances in the water.

The Aviemore district is tremendous country for the naturalist. Standing massively on the southern skyline is the great tableland of the Cairngorms, the most extensive area of high mountain in all Britain, with four separate peaks exceeding 4000 ft. and carrying on its lower slopes the biggest single tract of native woodland, largely of Scots pine.

The scene has changed greatly since Seton Gordon wrote his evocative books depicting the Cairngorms as a wild and untrodden land of forest and mountain, accessible only to the hardier sorts of walker. The mountain road to Coire Cas, the ski-tows and chair lifts on Cairngorm and the recreational developments around Loch Morlich have destroyed the wilderness character of this fine area to the east of the Lairig Ghru, whilst recently-cut Landrover tracks to the tops elsewhere have greatly detracted from the atmosphere of the mountains. The lover of quiet places will hasten through Aviemore nowadays. Yet the beauty of the area and its wildlife is otherwise unchanged, and those prepared to venture away from the tarmac can soon be on their own.

Wander into the pinewoods and you can soon be lost to the outside world. They are large enough for one to capture, for a moment, the sense of solitude of the vast northern forests of the Boreal regions. The fine pinewoods of Rothiemurchus, Abernethy (*Fig.40*), Insh-riach, Invereshie and Glen Feshie are amongst our national treasures. Glenmore has been somewhat spoiled by the planting of exotic conifers, as well as by tourist developments.

The native Scots pine is a most beautiful tree. Look carefully at each one as you pass; notice the marvellous and endless variety in form, and the range in texture and colour of the bark. Form depends partly on how closely the trees grow to each other. Where they are densely grown, through planting or the copious regeneration of seedlings naturally, the older trees have tall, slender trunks, branching little below and carrying the foliage in a high canopy. Such woods are shady places at ground level, and the darkest of them have little other than dense carpets of moss beneath. Where more light filters through the canopy, thick growths of bilberry and cow-berry grow through the moss, and as the wood opens out, heather enters in increasing quantity. As the trees become more widely spaced, so they are able to express their in-herent variety of form. The forest thins out in many places to a savanna-like heather moor-land with scattered pines, and locally it is interrupted by patches of wet peat bog occupying poorly drained hollows and channels. The pines seed onto these bogs, which are often quite thickly grown with little trees that belie their age, for many are runts whose growth is held in check by the wetness of the ground.

The trees decrease gradually in stature with altitude, but only in a very few places do they show a natural upper limit. The best surviving example of this feature is at 2100 ft. on Creag Fhiachlach, a spur of the Cairngorms facing Aviemore, where a thicket of gnarled and dwarfed little pines, mixed with junipers, gradually merges into the open heather moorland (*Fig. 8*).

In many places where the pines are not too densely grown, junipers form a patchy under-scrub, or occupy open glades and gaps within the forest. The mixture of the two gives a more

Fig. 35 Tufted saxifrage – a very rare Arctic plant of high rocks on a few mountains.

Fig. 36 Alpine hawkweed – a particularly handsome, hairy form of this large group of plants, growing on a high rock ledge with northern crowberry and woolly fringe moss.

Opposite
Fig. 37 Cowberry and Crowberry – two northern dwarf shrubs of woodland, moorland and mountain. The crowberry here is the red-stemmed common form.

Fig. 38 Harebell – the Scottish bluebell, growing widely along roadsides and in open ground generally up to moderate elevations.

Fig. 39 Woolly willow – a rare and beautiful small shrub of a few elevated mountain cliffs, out of the reach of sheep and deer.

Fig. 40 *Abernethy Forest, Inverness-shire – one of the great pinewoods clothing the lower slopes of the Cairngorms, cut by a rushing mountain stream.*

natural and attractive appearance to the woodland (*Fig. 3*).

Some of the biggest trees are quite magnificent, and it would be fitting that they be cherished as ancient monuments, to fade out eventually in extreme age. There are pines hereabouts of 300 years in age. What scenes these old warriors must have witnessed during their long lives. And even in death, they remain splendid, dignified relics.

The pinewoods are not rich in flora, but they have some plants of special distinction. One of these is an orchid, the creeping ladies' tresses, which sends up its spikes of cream-coloured florets in July, and is to be found only in association with the Scots Pine. The lesser twayblade of the heather moors is here, too, and there is plenty of chickweed wintergreen, a a dainty little relative of the primrose. The common wintergreen, with shining, rounded leaves, is frequent, but there is less of the serrate wintergreen, which has more pointed leaves and the little bells of its flowering spike all turned to one side. Still more careful search has to be made for the beautiful twin flower (*Fig. 33*), whose creeping stems are difficult to spot when the plant is not in bloom, Though mainly a pinewood species, it occurs also under birch and occasionally amongst heather on open moorland.

The most elusive of the pinewood specialities is the single-flowered wintergreen or St. Olav's candlestick (*Fig. 42*). It occurs in scattered localities in the eastern Highlands, but its distribution is a mystery, for the habitat appears to be in no way unusual, and is so widespread. The flowers are relatively large, flattened and mostly inclined downwards, as though the plant prefers shyly to hide its face. Most of these plants thrive best where the bilberry, cowberry and heather of the pinewood floor are not too dense, and there are carpets of moss, mixed with needle litter, cones and fallen twigs or branches.

Some common woodland plants are plentiful all over these pinewoods, among them the common cow-wheat, hairy woodrush and wavy hairgrass. Ferns are conspicuous in many places, bracken being generally distributed and the hard fern abundant. Mountain fern, beech fern and oak fern are more patchy, but quite common overall (*Fig. 15*).

The mosses—soft, golden yellow and luxuriant—are worthy of notice. One particularly fine species has a feathery appearance, with its silky branches ending in golden curls. It has no English name but rejoices in the scientific one of *Ptilium crista-castrensis*. And the tiny, fern-like fronds of Thuidium form a green lace-work over parts of the forest floor (*Fig. 10*). In damper places, as on steep, shady slopes, there are large cushions of bog moss, variously coloured from green through delicate pink to deep red.

Oak is sparse and patchy in occurrence on Speyside, but birch grows with the pines in many places, and mixed woods are widespread. There are quite extensive woods dominated by birch in the Spey valley, but mainly away from the granite and on the soils formed from the softer Moine rocks. Fine birch woods occur on the lower slopes of the Monadhliath around Aviemore, at Kincraig, and along the Spey Valley to Ruthven and Glen Tromie. The birch here is mainly the pendulous kind, with deeply corrugated bark on the older trees. Juniper is again abundant in many of these woods.

The flora has a good deal in common with that of the pinewoods, notably in the abundance of chickweed wintergreen, hairy woodrush, wavy hair grass, bilberry, cowberry and mosses. But on the more fertile soils there is greater variety. Heath violet and primrose are often abundant, along with other small herbs such as self-heal, barren strawberry, wild strawberry, germander speedwell, common speedwell, celandine, meadow buttercup and wood sorrel. Sheep and deer graze through these woods, and crop down plants which would form a tall herbaceous growth of the kind usual in many sub-alpine Scandinavian birchwoods. These bigger herbs include the wood cranesbill, globe flower, melancholy thistle, water avens, meadowsweet and marsh hawk's beard, all of which may be seen growing tall and flowering well in a variety of ungrazed situations; on roadside and railway verges, river banks, hay meadows, rough field corners, and the ledges of elevated mountain cliffs. They form a handsome group which one meets with pleasure as old friends in various parts of the Highlands, though the wood cranesbill becomes uncommon north of the Great Glen.

Although it lies just outside the Highland and Islands, there is a most interesting birchwood on the hillside of Morrone above Braemar, on the Dee side of the Cairngorms. It is a high level wood of the downy birch with a dense underscrub of juniper, but opening out here and there into grassland and little marshes with stony flushes where water wells up to the ground surface. Amongst the junipers and in the grassland are mountain plants such as alpine cinquefoil, viviparous bistort, alpine ladies' mantle and northern bedstraw, whilst the wet ground has yellow saxifrage, Scottish asphodel, three-flowered rush and alpine rush. This Morrone wood shows a remarkable similarity to sub-alpine birchwoods in certain parts of south-west Norway, but appears to be the only one of its kind now remaining in Britain.

All these woods, whether of pine or birch, are grazed by red deer, and many of them by sheep as well. In consequence there is often a dearth of young trees, for the seedlings and saplings are browsed and stand no chance of growing beyond the vulnerable early stages. Unless the animals are greatly reduced in numbers, or excluded by means of fences, such woods will eventually become moribund and disappear. For even the longest-lived trees do not go on forever, and have to be replaced by new generations if woodland is to survive.

Yet in other places the natural regeneration of both pine and birch is good, and there is an abundance of young, healthy trees growing up in open places within or around the existing older woodlands. Trees do not usually regenerate successfully under the shade of a closed canopy, but as soon as gaps allow more light to enter, conditions become more favourable for the establishment of seedlings.

Many pinewoods are managed commercially, involving the rotational felling of blocks and replanting, though sometimes the trees are allowed to seed in naturally. Large woodlands are a valuable asset which have to be managed carefully by their owners to yield a return. Provided there are no excesses of clear felling, and that replanting is with the native trees, there need not be any appreciable loss of amenity or natural history interest. Indeed, much of this interest stems from the past regimes of management, which have produced such diversity in age and stocking density within each forest. The conservation of these forests, especially the pinewoods, should include the leaving of certain stands of particularly attractive, old trees to follow their natural course of maturation, decline and regeneration.

The open heather moors into which the pinewoods merge are the prevailing type of vegetation on the slopes of the Cairngorms up to around 3000 ft. The sequence noted in the Drumochter hills occurs here also, with alpine heaths, in which heather is mixed with other dwarf shrubs, taking over at the higher elevations. Bearberry is especially abundant here. There is also much dwarf azalea, studded with its tiny deep pink blossoms during June.

On some of the spurs above Glen Feshie these dwarf shrub heaths have great quantities of lichens of the 'reindeer moss' type, colouring the ground grey in places. It was, in fact, the abundance of lichens in the Cairngorms which encouraged the introduction of a small herd of reindeer to the area, for these curious plants are the principal food of the animal in its Arctic home. The Cairngorm reindeer eat a good deal of sedge and grass besides lichens, and seem healthy enough on this varied diet.

The 'peat alpines' are all well represented here, and there is a southern outpost for the alpine bearberry. Deer sedge is more prevalent than cotton grass in the bogs here, and is also abundant with mat grass in places where snow lies late.

The Cairngorms have a much larger extent of late snow influenced vegetation than any other mountain system in Britain. The high corries and sheltered slopes of these massive granite hills collect huge snowfields during winter and early spring, and in an average year, several large patches of snow last the whole year round. Snow cover has actually increased during the run of cold, backward springs since 1960, and in some seasons the ground at over 3000 ft. has had a truly Arctic appearance at the beginning of June (*Figs. 18 and 43*).

This is terrain where climate has to be taken seriously, for conditions can change from the benign to the ferocious with great rapidity. The winter snow-blizzards are said by experienced mountaineers to be as bad as anywhere in the world. During spring snow-melt, some of the higher corries regularly experience considerable avalanches, and some of the high lochans contain ice for eight months in the year.

The dark green to blackish carpets of moss and liverwort associated with really late snow cover are extensive, and there is particular abundance of associated alpines such as Sibbaldia, alpine ladies' mantle, least cudweed, spiked woodrush, and alpine tufted hair grass. Species growing in the cold mossy springs and moist rocks are well represented, such as starwort mouse-ear, alpine speedwell, the mountain form of thyme-leaved speedwell, alpine foxtail grass, alpine timothy grass, Highland saxifrage, alpine willowherb, starry saxifrage and hare's foot sedge. Block litters in many places are densely grown with the alpine lady fern and parsley fern, which seem to depend on the protection from extreme frost which, strangely, is given by the winter mantle of snow.

Fringe moss heath is much less continuous on the Cairngorms than on the Drumochter summits. On the high windswept ground where snow lies but thinly, the vegetation cover tends to be in mosaic with bare gravel and coarse sand, whilst many areas are covered with block litter or thickly strewn with deeply bedded boulders. The wiry three-pointed rush grows in tufts amongst the fringe moss mats, and above 3500 ft. it increasingly becomes the most abundant plant. The high tops have vast quantities of this rush and its tussocky growths glow with the richest of rust-brown when lit by the evening sun in late summer. Where snow lies rather more

Fig. 41 Scots pine forest at Loch Garten, Abernethy Forest, Inverness-shire, showing understorey of juniper and field layer of bilberry and heather.
Extreme right
Fig. 42 St. Olav's Candlesticks or single flowered wintergreen, a rare plant of north-eastern pinewoods.

deeply there are large patches of very short mat grass and lichens, forming a more continuous plant cover, and these change in places to dense growths of rigid sedge, moss and lichens. On the dreary waste of the Great Moss above Glen Feshie, such communities are mixed with patches of shallow blanket bog, here at its highest elevation.

The high Cairngorm summits have a few notable plants, such as the curved woodrush and wavy meadow-grass, but their corries are more productive. Besides the plants especially associated with late snow-lie there are local alpines such as northern rock-cress, Arctic mouse-ear and a fine array of hawkweeds (*Fig. 36*). The crags are mainly of acidic granite, but occasional patches of lime-bearing rock produce surprises, such as the very rare tufted saxifrage (*Fig. 35*).

Towards the flanks of the massif, where the granite changes to schistose rocks or even limestone, as in Glen Feshie to the west and Inchrory on the east, there is a quite spectacular increase in richness of the flora. In a few places the discerning botanist will recognise a strong similarity to the botanical character of the Ben Lawers range, from the abundance of the common alpines of calcareous schist, and the presence of many of the rarer ones. There are choice plants in the mountain avens, net leaved, whortle and woolly willows, rock speedwell, rock whitlow grass, rock sedge, black sedge, alpine saxifrage, alpine cinquefoil and mountain bladder fern.

A remarkable schist hill on the Dee side of the Cairngorms has a colony of the very rare alpine milk-vetch, growing mostly in short turf. None of the good Cairngorm ground can quite match the marvellous variety in alpine plants of the Clova Mountains east of the road from Braemar to Glen Shee. The corries of Glen Doll, Corrie Fee, Caenlochan and Glen Callater perhaps rate equally in importance with Ben Lawers itself, and have some species unknown in Breadalbane, such as yellow oxytropis, alpine sow-thistle and oblong woodsia. The granite mass of Lochnagar forming the northern end of this massif has many of the distinctive Cairngorm plants too.

Speyside has yet other interest to offer besides its mountains, pine and birchwoods, marshes and lochs. For those who feel inclined to expend the minimum of energy, the low ground around Aviemore and Boat of Garten is quite rewarding florally, and even a saunter around the by-roads will produce much of note. There is some arable land, but a good deal of rough pasture and moor, most of it once under trees, and lying on deposits of glacial sand and silt. Roadside verges here have an abundance of ladies' bedstraw, harebell, wild thyme, bearberry, bird's foot trefoil, beautiful St. John's wort, and locally the common rockrose. This last is a plant which botanists from the south will associate with chalk and limestone, but here it grows on soils containing little lime.

Some dry, sandy moraines between Boat of Garten and Nethy Bridge have a most intriguing flora, consisting of a mixture of plants typical of chalk and limestone grasslands, and others belonging to quite acidic northern heath. In the first group are rockrose—in profusion—thyme, kidney vetch, mouse-ear hawkweed, burnet saxifrage, purging flax, eyebright, bird's foot trefoil and red clover; whilst in the second are bearberry, cowberry, mountain everlasting and mat grass. Some of the dry heather ground hereabouts has vast quantities of bearberry, and in places many of the plants just mentioned also appear, together with intermediate wintergreen, petty whin and bitter vetch. The riverside shingle beds and sandy banks are also quite rewarding places to search. Many of the same plants appear here too, and there are a few of the alpines, sprung from seed brought down by the water from higher ground.

The most spectacular plant here is the Nootka lupin, large, handsome and seemingly out of place. It is indeed not a true native, but has been growing wild by these Highland rivers, the Spey and Dee, for a very long time.

To the west of Badenoch, the Laggan road runs between high mountain ranges, Creag Meagaidh to the north, Ben Alder and the hills of Ardverikie Forest to the south. There is some remote deer forest country in here, and large areas of planted pinewood. Botanically,

Fig. 43 Cairngorm plateau – the most alpine mountain table-land in Britain, with long-lasting snow patches mixed with freshly fallen snow in June.

Fig. 44 Cloudberry – a widespread and often abundant plant of the high mountain bogs and damp heaths, growing here amongst heather.

Fig. 45 Moss campion, or cushion pink – it grows in profusion on mountain cliffs, and on the high stony summits of the northern Highlands, but descends to sea cliffs in places.

Fig. 46 Purple saxifrage – this early flowering Arctic-alpine belongs to open habitats rich in lime.

Fig. 48 Wild rose – a familiar wayside shrub of the lowlands, often with deeply coloured flowers.

Opposite
Fig. 47 Mountain avens – one of the few dwarf shrubs of calcareous mountain soils, growing in great profusion down to sea level in the north of Sutherland.

Fig. 49 Snow cornice and soil creep terraces near the summit of Sgurr Mor Fannich, Ross-shire.

Fig. 50 Melancholy thistle – probably the original 'Scotch thistle', ranging from low-lying meadows, roadside verges and river banks to cliff ledges at 3,300 feet.
Fig. 51 Bogbean – one of the common plants of moorland pools and lochans, shallow loch edges and spongy swamps.

Fig. 52 Zigzag Clover – a beautiful plant of the lowland meadows and waysides.

although these are quite good hills, there is rather little which cannot be seen on the Cairngorms or the mountains of Argyll. Most of the widespread alpines grow on the Laggan —Loch Ericht hills, and some of the rarer ones, too. Acidic rocks have a few rarities not yet mentioned, such as Highland cudweed and a scarce form of alpine lady fern, and limestone outcrops in the Ben Alder range have some of the Ben Lawers 'specials', whose discovery within the last 20 years suggests that the possibilities of this wild district are not yet exhausted.

Still farther west, beyond Loch Treig and the railway, lie the high peaks of the range which culminates in Ben Nevis (4406 ft.), the highest summit in Britain. There is more interesting ground in the high corries here, and when exploring untrodden ground in 1956, Eddie Blake had the thrill of finding the three rarest British saxifrages, the Highland, drooping and tufted, all growing together in a secret gully. Ben Nevis is the other mountain where snow usually lasts the whole year round, and the snow patch flora is well represented on its higher levels.

If, instead, we head eastwards from Aviemore, the elevation of the uplands falls steadily and there is a drier district of heathery grouse moors, rather limited in the variety of their flora, but superbly colourful during August when the heather is in flower. There are pine and birchwoods, quite extensively in places, and the broad valleys of the Spey and Findhorn rivers have some attractive country; but the scene becomes more agricultural with approach to the coast. The southern side of the Moray Firth is interesting, in that a number of plants reach their northern British limits here.

Most of them are plants of warm, dry lowlands which simply become restricted to the coastal belt here; for example, viper's bugloss, marjoram, hoary plantain, musk thistle, pendulous sedge, jack-by-the hedge, weld, wild mignonette, meadow saxifrage, orpine, birdsfoot, rough chervil and bloody cranesbill. The huge sand dune complex of the Culbin Sands is now largely afforested, but unplanted areas still remain, and carry a typical sand-hill and strand-line flora, with plants such as marram, lymegrass, sea rocket, sea purslane, restharrow and ragwort. Northern species include oyster plant, coral root, mountain everlasting, chickweed wintergreen and, in the pine woods, creeping ladies's tresses. And around Inverness may be seen a salt marsh flora, with sea aster, sea milkwort, sea plantain, Gerard's rush, thrift and scurvy grass.

The coastal belt between Inverness and Elgin is perhaps the part of the area with the greatest affinity to the Lowlands of Scotland. It is quite heavily wooded, largely with planted forest, but there is a good deal of quite rich farmland with the kind of roadside verge, hedge, pasture and arable weed flora which may be seen in many agricultural districts. The white scented may blossom of hawthorn hedges and the beautiful pink clusters of wild roses (*Fig. 48*) will strike a familiar note to visitors from the south. But not far inland the heathery foothills rise gently towards those great central mountain ranges which are the real Highlands.

The Northern Highlands

The mainland of the Highlands north of the Great Glen has some magnificent country, and in many ways is the finest part of the whole area. It is a large and geographically complicated district, with a great many mountain systems, lochs and glens and a coastline of enormous length, which would require a book on their own to do them justice. But it has rather few plants which cannot be seen in the Central and South-west districts. This brief scan of its botanical interest will thus be directed towards the highlights, beginning just over the Ardgour ferry near Fort William and ending in the Black Isle of Easter Ross.

The two peninsulas of Morvern and Ardnamurchan, separated by the deep, sheltered sea inlet of Loch Sunart, have a great deal of basalt, which is quite a good rock for plants, since it often contains moderate amounts of lime. In the north of Morvern the two hills, Beinn Iadain and Beinn na-h-Uamha, have basalt cliffs and screes with a good flora which includes the mountain avens, alpine saxifrage, purple saxifrage, northern rock-cress, moss campion and rose-root. There is a colony of the rare Highland sandwort, an inconspicuous little plant which occurs in a few widely scattered localities, all in the western Highlands. On the coast facing the Sound of Mull are attractive woods with much ash and hazel and these have the lovely sword-leaved helleborine (*Fig. 61*), one of our finest native orchids, with an extremely local distribution. The fertile soils of these woods and cliff slopes above the sea have an abundance of plants such as primrose, sanicle, dog's mercury, enchanter's nightshade, wild strawberry, woodruff, bugle and wood false brome grass. Primroses are often in great quantity in quite open grassland in this area and make an extremely fine show in the spring.

Ben Hiant on the north side of Loch Sunart is another basalt hill with quite an interesting flora, and on rocks farther west is an isolated colony of the rare and curious forked spleenwort fern, with its tufts of stiff little sword-like fronds. The Ardnamurchan area has special interest in the occurrence of two plants, one an orchid and the other a relative of the lily family, which both belong really to North America, in that they are widespread there but in Europe are confined to Ireland and a few parts of Western Scotland. How they come to have these outposts on the eastern shores of the Atlantic is a mystery, and has kept plant geographers arguing for a long time. The drooping ladies' tresses is known especially from around Acharacle, where it grows on shallow peat, often in rather disturbed vegetation, rather than on undamaged bog. The unusual pipewort is an aquatic plant which has recently been found in shallow moorland pools towards the tip of the Ardnamurchan peninsula.

The sheltered confines of Loch Sunart are hallowed ground for the specialists in lichens, mosses and liverworts. The combination of mildness of climate, extremely heavy rainfall, and freedom from atmospheric pollution make this one of the richest parts of Europe for these groups of plants, and the lichens are outstandingly important.

They grow in a wide variety of habitats, from shore level upwards, but the native woods of oak, birch, ash, wych-elm, hazel, rowan and holly are particularly productive. The woodland flora is in general much the same as that in southern Argyll, but the number of liverworts and lichens is greater. Many of the older tree trunks are thickly grown with large and conspicuous lichens such as the tree lungwort and its relatives, and the rocky floors of the woods have luxuriant golden-green carpets of mosses and leafy liverworts. The two filmy ferns and hay-scented buckler fern flourish in some of these woods and their rock-bound ravines. Ferns are in general a conspicuous part of the vegetation. Stream gorges here have a few sub-alpine plants, such as yellow saxifrage, alpine ladies' mantle, green spleenwort and mountain melic grass.

Symers MacVicar of Invermoidart was both the local family doctor and one of our leading students of the liverworts who first revealed the outstanding richness of this district for these plants. His book on the British liverworts is still the standard text on these interesting and ancient plants.

Much more recently, the still more remarkable lichen flora of the Sunart area has been investigated in detail and with great patience by Francis Rose and Peter James, who found many rare species, including some unknown elsewhere in the British Isles. This varied assemblage of lowly plants has survived the vicissitudes of great shifts of climate over tens of thousands of years, but its survival is now in some doubt because human activity, in its relentless, all-pervading way, is causing profound change in the habitat.

There has been quite extensive planting of alien conifers in the area but it is particularly important that the remaining native woods are preserved, for the interest of the trees themselves and the rich flora they support. It is instructive to go from one of these native woodlands, with its marvellously varied assemblage of plant life, to a conifer plantation, especially of Sitka spruce, so dark, dreary and often almost devoid of vegetation other than the trees themselves.

Whilst there is no denying the economic necessity of utilising poor hill land for growing timber, it is always sad when the usually relatively small areas of land with really high wildlife value have to be sacrificed to meet this need. Some of the native woodlands are the remnants of forests which have developed their wealth of associated plant and animal life over thousands of years. When such a wood is clear felled, many of the creatures which depended on the trees simply vanish rapidly, and even if a cover of the same kind of trees is eventually restored, many of these woodland denizens do not come back. For their powers of spread are often small, and when the nearest survivors are some distance away, the chances of their jumping the gap are slight.

When the trees are replaced by quite different species, which grow up to give contrasting conditions of shade and soil, only the most tolerant and prolific of the original woodland plants have a chance of success. When important wildlife areas are known, there is no excuse for ruining them virtually forever while there remain great expanses of quite ordinary ground which would suffer much less loss by being 'coniferized'.

The mountains surrounding Loch Shiel, northwards of Sunart, are steep but not particularly exciting. They have many of the more common alpines of poor soils, such as alpine ladies' mantle, dwarf azalea, least cudweed, least willow, spiked woodrush and three pointed rush. They have taken a good deal of punishment by sheep and deer, so that the dwarf shrubs are poorly represented.

One rocky summit in this district, however, has the great distinction of being the habitat of an Arctic plant which was unknown in Britain before 1951, Diapensia. In that year ornithologist C. F. Tebbutt had the good fortune to discover a fine colony on ground which looks very little different from that on a great number of other high summits in the western Highlands. Its presence here is another of those intriguing mysteries of plant geography which defy explanation and lead one to suppose that some strange quirk of chance must have been involved.

Between the road from Loch Eil to Mallaig and that from Glen Moriston to Kintail lies one of the least trodden tracts of mountain land in the whole of Britain. It contains the particularly remote district of Knoydart, deeply indented by the fjord-like sea-lochs of Loch Hourn and Loch Nevis and containing the abrupt cone of Sgurr na Ciche (3410 ft.), and the more massive Ladhar Bheinn, carved into a magnificent north-eastern corrie. There may still be noteworthy botanical finds to be made hereabouts, but the ground is not particularly promising, since rather acidic Moine rocks prevail.

The country north of the Cluanie–Glen Moriston road is better, for the hills are much higher and there are many summits over 3500 ft. These ranges are best approached from the Inverness side, via the three great glens of Affric, Cannich and Strathfarrar which drain eastwards and converge to form the Beauly River. This district contains the other main tract of native Scots pine forest, though much of the old timber has been taken out in recent years and, despite extensive replanting with the same species, the pinewoods here tend to have a rather managed appearance. The best pinewoods, well mixed with birch, are those in Glen Strathfarrar, which has a private road, though the key will usually be made available to enquirers (*Fig. 65*).

The pinewood flora mentioned for the Speyside woods occurs here, though twinflower and single-flowered wintergreen are evidently unknown. The main lochs of this district, in common with those over much of the northern Highlands, have been modified in the interests of hydro-electric power generation.

Many have been enlarged considerably by dams which have raised their levels, and some have been created by damming rivers. In either case, their artificial shore-lines are virtually devoid of botanical interest. Often there is an unsightly draw-down zone which at low water appears as a band of bleached stones and sand, or disintegrating peat. Such conditions are unfavourable to colonisation by aquatic plants which under more stable conditions and in time, would begin to appear around the shallow edges. Most of the plants which once grew along these loch margins, however, still can be seen around the edges and in the shallows of the many smaller, undisturbed lochs in the area. They consist mainly of the species described for Rannoch and the Spey valley, with few additions, for the lakes of the Highlands are mostly fed by fairly pure water and so are poor in dissolved nutrients on which plants feed. Their flora is therefore rather limited in variety and tends towards a marked uniformity from one site to another.

The lower slopes of the hills have a good deal of shallow bog or wet heath, with mixtures of heather, deer sedge, purple moor grass, bog myrtle and bog mosses. Little sedgy and rushy marshes here and there show the influence of lime-rich drainage water by the abundance of yellow flowers of the meadow buttercup and autumnal hawkbit. In late summer they are often dotted with the creamy-white blooms of the grass of Parnassus—no grass really, but a saxifrage ally.

Steeper, dry ground has quite extensive areas of heather, and the dwarf heather community of the alpine zone occurs here at 2200–2400 ft. As usual, dwarf azalea, crowberry and bearberry are abundant amongst the prostrate heather, but there is also a good deal of the alpine bearberry, which differs from the other in having leaves with deeply impressed veins, and in being deciduous instead of evergreen. Before the leaves of alpine bearberry die off, they turn a beautiful shade of rich wine red, imparting rare colour to the ground where the shrub grows.

Several others of the mountain plants produce in their modest way a fine array of autumn colours. The dwarf cornel of the heathy slopes is another which turns a brilliant red, whilst on the cliff ledges, the larger and fleshier leaves of rose-root and mountain sorrel produce a range of yellows, browns and reds. Even the grasses and sedges contribute to the colour scheme with their more subtle tones.

Beyond the limits of dwarf shrub heath there are large expanses of fringe moss heath along the high ridges, containing abundance of rigid sedge, least willow, alpine ladies' mantle and moss campion. The hills at the head of Glen Affric reach 3877 ft. on Carn Eige, and there is a great deal of the various kinds of vegetation influenced by long snow cover.

Bilberry—crowberry, mat grass and tufted hair grass communities are extensive, and on some high slopes are quite large areas covered with golden-green mosses. The latest snow patch types, with dark felt-like moss swards, grown with Sibbaldia, least cudweed, spiked woodrush, least willow and alpine willowherb are well developed; and here and there are cold springs covered by bright apple-green and frothy-looking cushions of moss. The alpine flora is not particularly rich, but there are a few rare plants of acidic soils, such as Highland cudweed, curved woodrush and starwort mouse-ear. Alpine ladies' mantle is in great profusion, and three pointed rush grows plentifully along the high ridges.

There is a good deal of unworked ground in these ranges. The woolly willow was found fairly recently in the hills towards Cluanie, and perhaps other good things will be discovered yet.

North of Loch Carron lie the imposing hills of Applecross, Coulin, Torridon (*Fig. 55*) and Loch Maree, formed mainly of two kinds of rock which are important in the north-west Highlands. The Torridonian sandstone is a dull dark red or even greyish in colour, and produces horizontally layered cliffs; whereas the Cambrian quartzite is pale grey to white, and forms summits which gleam in the sun as though clad with fresh snow. The sandstone is mainly an acidic rock, but contains a little lime in places. Its cliffs are often very wet and so have a fine development of stable ledge vegetation.

Beinn Bhan in Applecross is one of the most interesting sandstone mountains, though its lower slopes are especially sterile and degraded by excessive grazing and burning. One of its eastern corries has a great sloping ledge about 250 yards long by 50 yards wide, cut off above and below by cliffs. It can be scaled from below, but not by deer or sheep, and its vegetation is a most dramatic illustration of the influence that these herbivores exert on their habitat (*Fig. 67*). The whole ledge is covered

Fig. 53 Mountain or rigid sedge, an abundant alpine of the high tops.

largely by tall and luxuriant beds of ferns, especially the alpine lady fern, mountain fern, broad buckler fern and male fern, growing to an average height of 2 to 3 feet. Where water seeps down the ledge and enriches the soil there are contrasting strips and patches of tall herbs such as marsh marigold, meadow buttercup, sorrel, water avens, valerian, angelica, meadowsweet, globe flower and marsh hawk's beard. Below the cliffs, on the grazed corrie slopes, are bare screes and patchy areas of tufted hair grass contrasting with the profusion of the ledge vegetation and suggesting most strongly how this is either destroyed altogether or completely changed under the attack of grazing animals.

Beinn Bhan and some of the other sandstone mountains such as Liathach, the Flowerdale hills and An Teallach above Dundonnell have an abundance of some of the more widespread cliff plants, such as roseroot, alpine saw-wort, mountain sorrel, alpine scurvy grass, globeflower, thrift and mossy saxifrage. Some of them have more local alpines such as northern rock-cress, Arctic mouse-ear and alpine tufted hair grass, but species which need much lime in the rock are mostly absent.

During the last few years, however, Ron MacBeath has made the discovery of tufted saxifrage on two widely separated Torridonian sandstone mountains. This plant was not previously known north of the Great Glen and all its other localities are on highly calcareous rocks. It is a true Arctic plant not found in the Alps.

The Beinn Eighe massif has a great deal of quartzite, spectacular in the poverty of its vegetation cover. Some of the upper ridges and corries appear at a distance to be almost devoid of plant life, though closer examination shows a sparse growth of the more widespread kinds. One area has an interesting alpine heath in which short heather is mixed with dwarf juniper, and both the bearberries are plentiful in places. Dwarf juniper is extremely vulnerable to fire and on many hills in the north west, its gnarled remains, bleached and long-dead but slow to decay, show it once to have been much more abundant. It has survived in quantity on Beinn Eighe because it grows in an open patchwork on quartzite scree, so that fires have been unable to run through the whole colony. The influence of lime-bearing rock on vegetation is nowhere better seen than on Beinn Eighe, where highly localised occurrences of calcareous mudstone amongst the quartzite produce a sudden lushness, green-ness and continuity of the plant cover amidst acidic heaths or even sterile stone deserts.

The lower slopes of Beinn Eighe above Loch Maree have an important native pinewood, though it was ravaged by felling when wartime need for timber was urgent. The Nature Conservancy Council are attempting to restore its former glory through both planting and natural regeneration of the trees, though this is up-hill work under the harsh climate of the north-west. There are other important pinewoods on the larger islands of Loch Maree, and there is another farther south at Shieldaig above Loch Torridon. The Shieldaig wood was swept by fire in 1975, with results which show only too well the disastrous consequences of a single act of thoughtlessness. These north-western pinewoods differ from those of the central Highlands mainly in the greater amounts of heather and bogmoss, and in the absence or scarcity of some of the characteristic herbaceous plants. On the opposite, sun-exposed south-west facing slopes at Letterewe above Loch Maree are extensive woods of oak, evidently of native origin, but with pines growing scattered on rocky bluffs where the soil is thin and acidic. The influence of soil fertility on woodland type is well shown in the Beinn Eighe woods, where the prevailing pine on quartzite gives way to birch on an area where the soil is extensively influenced by downward seepage from calcareous mudstones.

On the coast, near the foot of Loch Maree are the celebrated Inverewe Gardens, now owned by the National Trust for Scotland. The secret behind the creation of these gardens was the planting of an effective shelter belt of trees, to give protection from blasting by the violent westerly and salt-laden winds from the Atlantic. Behind this barrier has been established a marvellous collection of tropical and sub-tropical plants, albeit of the hardier kinds which could withstand the still rather cool conditions of the site. Osgood Mackenzie, who founded the gardens, described how in the mid 1800s, peaches ripened annually on the walls at Conon House near Dingwall and colonies of wild bees in Wester Ross produced honey in sufficient quantity to be worth collecting.

The climate took a turn for the worse during the second half of the 19th Century, and the colder conditions put an end to the peaches and bees. Northern birds such as the snow bunting nested more regularly and widely during this colder period. The opening years of the present century saw a return to somewhat milder conditions again, only to be followed by yet another cooling after 1940. The present period is one of cold and backward springs, with massive amounts of snow often lying on the higher hills in June. These minor ups and downs of temperature represent the ripples on the waves of the much larger fluctuations of climate which take place over thousands of years. During any one person's lifetime the observable effects on plants and animals are small, but there is a slow and steady shift—a slight cutting back of range of one species here and a little expansion of something else there—which accumulate in the longer term into major trends of change.

The deforested lower moraine ground of the north-west has a great monotony in flora. For mile after mile there is the same mixture of deer sedge, purple moor grass, cotton grasses, rather sparse ling and bell heather, cross-leaved heath, patchy bearberry, bog asphodel, fir clubmoss, bog moss and woolly fringe moss. The flowers of lousewort, heath spotted orchid and mountain everlasting (*Fig. 56*) add a little colour during early summer, but there is a general poverty in botanical interest.

Wet seepage areas and stony flushes offer rather more variety. The black bogrush, with pale, wiry tussocks and very dark flowering spikes is a characteristic plant of such places in the north-west, and the English sundew is often especially robust in this habitat. Richer sites have yellow saxifrage, yellow sedge, few flowered spike-rush, and broad-leaved cotton grass, which has shorter, flatter and yellower green leaves than its more common relatives. The common butterwort is abundant in these damp spots and in its company there may be found the less plentiful pale-flowered butterwort with smaller, more delicate leaves and pale rose-pink flowers. A still rarer plant of the open flushes in the Loch Maree area is the curious marsh clubmoss, an inconspicuous inhabitant of wet heaths mainly in England. It has become quite rare, partly through reclamation of these places, but seems always to have been an extremely local plant in Scotland.

The higher mountains of Wester Ross are on the whole much less rich in alpines than those of the schistose hills of the central Highlands. The quartzite and sandstone ranges are especially limited in flora, though it was on one of the Torridonian sandstone hills that the exciting discovery of Norwegian mugwort in Britain was made in 1950 by Sir Christopher Cox.

This strange little plant grows in some abundance on a windswept, stony fell-field plateau below one of the higher summits. It has since been found on two other tops in the district, growing inconspicuously in fringe moss carpets with other alpines such as moss campion and alpine ladies' mantle.

The second locality was found by Humphrey Milne-Redhead, a hardy Galloway country doctor who travelled big distances and sometimes slept rough on the hill, as on this occasion, to work remote ground. This is one of the important British mountain plants, for it is found in a few places in Norway and the Ural Mountains of the USSR, and nowhere else in Europe.

A few outcrops of limestone or calcareous mudstone greatly enhance the variety of the mountain flora in Wester Ross, but they are extremely localised, the best being in the hills above the head of Loch Kishorn, where there is mountain avens, whortle willow, holly fern and many of the more widespread lime-loving alpines. The higher mountains of Lewisian Gneiss are quite good in places, notably in the wild and precipitous hills of the Letterewe and

Fisherfield Forests to the north-east of Loch Maree.

But the richest of these north-western mountains are those formed of rocks of the Moine Series, particularly in the area around the Dirrie More road from Garve to Ullapool. Here, in the remote deer forest country of the Beinn Dearg—Seana Bhraigh massif to the north, and the Fannich hills to the south, there is the closest approach north of the Great Glen to the floral richness of the Ben Lawers range and the Clova mountains. The rock in many places is calcareous and friable, and there is a mica-schist superficially resembling that of the Dalradian Series farther south. Veteran plant hunters Ted Wallace and Bob Mackechnie were the first to open up the botanical riches of these sequestered hills, in the years after World War II (*Fig. 57*).

The approaches to these little-trodden hills are boggy, and the dwarf birch grows quite plentifully in places, often along with the alpine bearberry, and sometimes the common bearberry as well. Dry shoulders and spurs with dwarfed heather have an abundance of both bearberries, but dwarf juniper is less plentiful than on certain quartzite hills. The high tops and ridges at over 2500 ft. have great areas of woolly fringe moss, but here diversified by a profusion of dense, flat patches of three herbs which we have already seen abundantly as plants of cliff faces, flushed slopes and, more sparsely, rocky summits. They are the mountain form of thrift, moss campion and mossy cyphel, which on these Ross-shire hills all grow in great abundance, the first two adding greatly to the otherwise drab colours of these high level moss heaths with their myriads of pink flowers (*Fig. 45*). On the numerous areas of stony fell-field which also occur extensively on these high watersheds there is a sparser vegetation with these same plants, but including also the three-pointed rush, Sibbaldia, least cudweed, mountain everlasting and spiked woodrush. Places where water seeps intermittently have a grassier vegetation, with a greater variety of small or dwarfed alpines, such as alpine meadow rue, Sibbaldia, alpine ladies' mantle, and non-flowering plants of alpine saw-wort, mountain sorrel and viviparous bistort.

Although the summit areas of these Ross-shire hills are smaller than those of the central Highlands, they are captivating, lonely places where one remains especially alive to the possibility of some new discovery. Many of them have surface features which give a vivid impression of the still active moulding of these mountains.

In places are 'giant's staircases' of terraced slopes which represent a gradual downwards slumping of the loose layer of stone and soil which has formed on the high tops since they were abandoned by the ice (*Fig. 49*). On flat ground are many areas where blasting by ferocious winds has stripped much of the former layer of soil, leaving a largely open layer of stones appearing as if steam-rollered into the ground. The finer material carried away by the wind has piled up as new deposits in sheltered places to windward, there to be reshaped yet again by water and frost into hummocks or eroded once more by the relentless wind. Snow lies long on some of these hills, and there is the same range of late snow-bed vegetation as was noted in the Affric–Cannich mountains. This is the most northerly area in the whole Highlands and Islands where such communities are well developed and some of the characteristic species, such as star-wort mouse-ear, are not known farther north.

The mountain flora, however, is again best represented on the cliffs and flushed slopes of the corries and spurs lying below the highest summits. There is a fine development of 'hanging gardens' on some precipices, with abundance of yellow, mossy, starry and purple saxifrages, moss campion, mossy cyphel, rose-root, mountain sorrel, northern bedstraw, alpine scurvy grass, alpine ladies' mantle, alpine saw-wort, northern rock-cress, Arctic mouse-ear, alpine meadow rue, viviparous bistort, holly fern and three-flowered rush. Broad ledges have profuse growths of tall herbs of upland meadow, such as globe flower, angelica, meadow sweet, water avens, marsh hawksbeard and great woodrush; and some moist cliffs have a good deal of downy willow. There is a fairly long list of rarer alpines, found only in a few places, among them the mountain avens, net-leaved willow, alpine

Fig. 54 Falls of Measach, Ross-shire – a fine example of a waterfall gorge, 200 feet deep, clothed with open woodland and richly grown with mosses, liverworts and lichens.

Opposite
Fig. 55 Ben Alligin, Ross-shire – one of the magnificent hills carved from the Torridonian sandstone, rising behind the sea inlet of Loch Torridon and with pinewoods clothing the lower slopes.

Fig. 56 Mountain everlasting – a common northern plant related to the edelweiss, sometimes with pink flowers and here accompanied by the yellow birds foot trefoil.

Fig. 57 Hills south–east of Loch Broom from Achiltibuie, Ross-shire – a wild, little known mountain country with a rich alpine flora. Ruined crofts near the shore suggest the rigours of farming hereabouts.

Fig. 58 Oyster plant – a beautiful plant of sand and shingle on northern shores.

Fig. 59 Loch Aline shore, Argyll – the zone of seaweeds between tide marks passes to a characteristic belt of wild irises above high water level. In other places native woods, mainly of oak, reach to the sea shore.

Fig. 60 Purple marsh orchid – a handsome wild orchid of the machair marshes and richer fens.

Fig. 61 Sword-leaved helleborine – a rare woodland orchid, here in the Appin district of Argyll.

Fig. 62 The barren moorlands of Harris – a land of bare gneiss, peat and water. The peat is locally cut for fuel, as in the near left in the picture.

Fig. 63 The Trotternish ridge from the Quiraing, Skye – the tremendous escarpments formed by massive landslips are an important habitat for alpines.

Fig. 64 The coast of Hoy, Orkney – the tremendous line of cliffs running north from the Old Man rises to 1,100 feet at St John's Head. The moorland behind the cliffs has mountain vegetation such as alpine dwarf shrub and lichen heath growing at unusually low elevations.

saxifrage, alpine cinquefoil, alpine mouse-ear, rock whitlow grass, Scottish asphodel, black sedge, russet sedge, hair sedge, two-flowered rush, chestnut rush, alpine meadow grass and glaucous meadow grass. Two rare plants of more acidic high level rocks occur sparingly, the Highland saxifrage and Highland cudweed, but the alpine tufted hair grass is fairly widespread. There is a general luxuriance and abundance of mosses and liverworts on the shady slopes and rocks, in keeping with the extremely wet climate. The hills of the north-west Highlands are an important area for these lowlier plants, especially the leafy liverworts, some of which grow here in great profusion yet have an extremely localised and highly fragmented distribution in other parts of the world. Some of them are very much rarer on the global scale than most of the mountain flowering plants and ferns.

To the north of Ullapool and the Coigach Hills, the west coast road crosses into Sutherland at Knockan Rocks. Sutherland has great character and attraction for the naturalist. In the west it is a wild expanse of knobbly gneiss country and bogland, liberally dotted with lochs of all sizes; and from this rocky and watery wilderness there rise isolated abrupt and strange peaks and ranges with romantic names. The rock formations are the same as those of Wester Ross, and the higher hills have a rather similar flora, with limited variety on quarzite and sandstone and greater richness on Lewisian gneiss and Moine schists. The most rewarding of the higher hills to the botanist are Ben More Assynt, some of the tops of the Reay Forest and Ben Hope. A few of the rare alpines of Ross do not reach so far north, but most of them are to be found in one place or another. The massive outcrops of limestone at Knockan and Inchnadamph are important botanical localities, with extensive cliffs, screes, grassland and flushes. Mountain avens is in great abundance, and a rocky plateau above the Stronechrubie cliffs has the largest colony of whortle willow in Scotland. The dainty white-flowered Norwegian sandwort is here and other notable limestone plants are rock sedge, Don's twitch grass, dark-red helleborine, holly fern and limestone polypody. The green-ness of the vegetation in these limestone areas, and the presence of rich meadowlands around the crofts and farms is in marked contrast to the sterility of the prevailing acidic moorlands.

Farther west, the low but rugged coastal ground around Inverpolly, Lochinver and Drumbeg has a good deal of birchwood, often on rocky terrain.

Birch is the only native tree of any importance in Sutherland, though oak of certain natural origin occurs in the birchwoods almost as far north as Scourie. Mosses, liverworts, lichens and ferns are again profuse in these north-western woods, but some of the more southerly species do not reach so far north. Wilson's filmy fern is in great quantity, and there is hay-scented buckler fern in a few places. Partly wooded rocks around Achmelvich have a good deal of the handsome upright bitter vetch, but this very local plant can also grow in meadows, on roadside banks or amongst heather. The Sutherland birchwoods do not extend to much more than 1000 ft. in elevation, and their distribution may always have been somewhat patchy.

Many notable northern or alpine plants may be found here without expending great energy in climbing high hills. One of the most interesting is the northern bugle, which belongs especially to Sutherland and Wester Ross. It is a plant mainly of the lower ground, and occurs quite frequently on rocky banks beside roads, but is much overlooked since many of its colonies scarcely ever seem to flower. The little rosettes of rather pale green and hairy leaves look rather like young plants of foxgloves or ragwort, but the flowering spikes are handsome and unmistakeable, with dense form and purplish, hairy bracts.

In this far northern corner of mainland Scotland, the cool, oceanic climate allows alpine plants to grow at remarkably low levels. Many species occur plentifully here fully a thousand feet below their lowest occurrences in the central Highlands. The same is true of whole communities. The effect of altitudinal depression of 'life zones' of plants and animals (it is shown also by birds such as the ptarmigan) becomes still more marked on the moorlands which abut the storm-swept coast. The alpine dwarf shrub heaths, with prostrate heather, dwarf juniper, dwarf azalea, alpine and common bearberry, least willow and northern crowberry occur down to less than 1000 ft. elevation in the Parphe, that desolate tract of moorland which ends at Cape Wrath.

The downwards descent of montane vegetation is still more remarkably shown on the exposures of limestone which run down to sea level at Durness, and on the associated areas of blown shell sand which occur here, as elsewhere on the north coast. The lime-rich soils here have vast quantities of mountain avens, giving a most spectacular display with a profusion of creamy white flowers in the short turf at the end of May. In its company there are scattered colonies of the rare rock sedge.

The blown shell-sand areas on the low headlands around Bettyhill to the east are famous plant localities. Along with a further profusion of mountain avens are purple saxifrage, moss campion, and that most beautiful of our northern plants, the purple oxytropis. The seaward fringe of sand-hills has a rather typical dune flora, but in damp slacks there is the uncommon and odd little sea sedge, with its short and curved flowering stems.

The coastal headlands have fringes of a characteristic maritime heath with short heather, crowberry, creeping willow, various grasses, tormentil, heath-spotted orchid, bird's foot trefoil, heath violet, carnation sedge and sea plantain. In places along the north coast of Sutherland and Caithness where salt spray influence is fairly marked, this community has large amounts of the attractive pale-blue spring squill.

Damp places here are also the habitat of the tiny but beautiful Scottish primrose, one of our most distinguished plants, in that it is now recognised as a distinct species found nowhere else in the world but along this coast and that of Orkney. (*Front Cover*)

The great seacliffs have lush growths of certain mountain plants such as roseroot and, more sparingly, mountain sorrel and alpine saw-wort, along with the more commonplace mixture of great woodrush, red campion, angelica, sorrel and bluebell, as well as the typical maritime group with thrift, sea campion, scurvy grass, scentless mayweed, sea plantain and red fescue. The lovage, a northern coastal umbellifer, sometimes used as a salad plant, is fairly common and purple oxytropis has a number of localities on the seacliffs.

In many places damp moorland with deer sedge and heather, or actual peat bog, runs almost to the crest of the coastal headlands, and stretches away inland in almost interminable expanses, flat or gently undulating, but rising eventually into distant peaks or ranges. From the bogland of A'Mhoine between Loch Eriboll and the Kyle of Tongue, the tall peak of Ben Hope with its great western escarpment stands out finely to the south, and the multiple, rock-crowned summits of Ben Loyal rise distinctively a little farther east. This is the beginning of a vast tract of blanket bog which stretches with few interruptions right across north-east Sutherland and much of Caithness. It is the largest expanse of peatland in Britain, and dwarfs even such areas as the Muir of Rannoch. These great flat bogs, or 'flowes', are the real wilderness, and a long walk across them can give a sense of almost frightening solitude (*Fig. 66*).

There are no more desolate places in these islands than some of the enormous Caithness flowes around the upper course of the Thurso River. Plodding wearily over their spongy surfaces for mile after mile, the solitary wanderer feels an utter loneliness. Although the ground is broken with varying frequency by pools or dubh-lochans of all sizes, the surface is otherwise featureless. These peaty tarns with their inky depths of water add to the sense of the sinister. The only landmarks are far distant, notably the strange and compelling skyline of the conical Morven and its satellites, standing commandingly above the flowe country, well to the south. Farther west, Ben Griam More and Ben Griam Beag are twin cones rising steeply from the boglands at the head of Strath Halladale.

This great peat-clad terrain has something of the dreariness of the Arctic tundra to many people. There is no denying that its flora is extremely limited, showing close similarity to

Fig. 65 Birchwoods, with a field layer of bracken, Glen Coultie, off Glen Urquhart, Inverness-shire.

Fig. 66 Bog bean dubh lochans on flowe under Ben Loyal, Sutherland.

that described for the Muir of Rannoch. In places where the bog surfaces have not been too badly burned there is an abundance of the two bearberries and dwarf birch. Bogmosses of many kinds carpet the undamaged surfaces, typically forming undulations of hummock and hollow. Where the ground is especially wet the myriads of bogbean pools and lochans are a distinctive feature. It is often noticeable that where the bog surface has perceptible slope, however slight, the pools are elongated, and are arranged so that their long axes are at right angles to the direction of slope. From a higher hill overlooking such a bog, it appears almost as though the bog surface were splitting and beginning to slide down the slope. Such is not actually the case, but the manner in which these pool systems have formed has long puzzled bog ecologists. Bog-bursts, with the semi-fluid mass of peat suddenly flowing sideways in tidal-wave fashion are of surprisingly rare occurrence through recorded history in Scotland.

In drier or more heavily burned areas, deer sedge has often become dominant on the boglands.

The peat has been cut locally for fuel, and some years ago an attempt was made to exploit these vast deposits on a commercial scale. A peat-burning experimental turbine for generating electricity was built near the remote Altnabreac station on the Lairg—Wick railway, in the heart of these desolate moors. It operated for several years but closed when the venture finally proved uneconomic.

Some of the drier areas of bog are being extensively afforested with conifers by the Forestry Commission, as a drive along the wild road from Lairg to Altnaharra will show. This last great wilderness is under pressure for development as the present economic circumstances of Britain cause people to look hopefully at remaining 'waste' land and consider the possibilities for exploitation.

Caithness is formed mainly of Old Red Sandstone, which along the coast has been cut by the sea into long ranges of vertical, flagstone-layered cliff, beloved of nesting seabirds. Dunnet Bay has a low shore fringed by sand dunes and backed by a large, flattish inland stretch of blown sand. This is quite good ground botanically, and the north-east corner of Caithness is rather more fertile and modestly floral than the flowe country to the south and west. This is farm and crofting land, with much pasture and marginal ground reclaimed from the moor. Some of the bigger lochs, such as Watten and Calder are fed by richer water than those of the boglands, and in places are mixed swamps with a variety of sedges and other fen plants. Native woodland is virtually absent, and there is a general bleakness to the scene. Good birchwoods appear at Langwell near the Sutherland border, and from there southwards, but Caithness is one of the most treeless parts of Britain.

The mountains of east Sutherland and Caithness are less rewarding than those of the west. Most of the more widespread alpines of acidic ground are to be seen and the two Ben Griams, surrounded as they are by great sweeps of blanket bog, have quite an interesting flora with a number of lime-loving alpines. The much higher Ben Klibreck, which stands so finely above Althaharra farther west, is a rather ordinary hill for mountain plants, but the spongy swamps by Loch Naver at its foot are the only known place in Britain for the string sedge.

The south-east of Sutherland has a good deal of unspectacular moorland on which heather is prominent. A plant new to Scotland came to light in this area in 1961—the rock cinquefoil, previously known in Britain only in two places in the Welsh Borders, where it is now almost extinct. A strong colony of this interesting species was found on rock ledges, and the discovery in 1976 of a second locality some miles away sets its native status beyond doubt. The rockrose reaches its northernmost limits near Golspie, and in pinewoods here is a famous locality for the single-flowered wintergreen.

The fine oakwood at Spinningdale is the most northerly in the country, but perhaps still more interesting is the large alder and willow swamp woodland which has developed quite naturally on the former estuarine sands behind the Mound. This area was sealed off from the sea by the building in 1816 of the artificial embankment across the upper part of Loch Fleet.

This north-east coast of the Highlands has some tremendous shows of gorse, always a plant to delight botanist visitors from the Continent, where it is extremely local. Whole hillsides glow yellow in places, and broom often adds to the effect, for it flowers around the same time, in late spring. The bird cherry also makes a fine display along many roadsides at the end of May, when the handsome spikes of white flowers show it to be quite a common shrub.

Strath Oykell above Bonar Bridge has a complex of flood plain meadows and marshes where many of the plants of northern fen can be found. The Oykell flows into the long estuary of the Dornoch Firth, flanked in places by mudflats, salt marshes and sand dunes. On the flats, profuse growths of eel grass are an attraction to wildfowl, and the salt marshes have some distinctive plants in Scottish scurvy grass, shore centaury, saltmarsh flat-rush and glasswort. The dunes of Morrich More have a rich flora, though there is an RAF bombing range here and access is limited. Southern plants such as purple milkvetch and spring vetch grow here, along with northerners such as a Highland form of autumnal gentian, mountain everlasting, lesser twayblade and sea sedge. On more acidic parts of the dunes there is a good deal of crowberry with the heather and an abundance of a low form of juniper.

To the south is the still longer estuarine inlet of the Cromarty Firth, now in process of becoming a major centre of industrial development. Its southern shore is formed by the Black Isle, a largely agricultural peninsula but with interesting pine-grown bogs. It is here that the alpine butterwort, a plant unknown in Britain today, is said once to have grown, before draining destroyed its habitat. This dry and sheltered area of the upper Moray Firth is the most northerly place in the country for certain plants, among them hound's tongue, great mullein, carline thistle, giant bellflower, moschatel, wall lettuce and greater reedmace. Rocky headlands bound either side of the narrow entrance to the Cromarty Firth, and have a noteworthy species in purple oxytropis.

The great whaleback mass of Ben Wyvis (3429 ft.) looms high behind the Cromarty Firth. There are fine corries, and snow patches lying until late in the year have the characteristic range of plant communities, but the micaceous Moine rocks which form Wyvis are here mainly acidic, and its montane flora is rather limited. Alpine foxtail grass and Highland cudweed are among its rare plants, and some of the 'peat alpines' such as dwarf birch, both bearberries, cloudberry, dwarf cornel and interrupted clubmoss are locally plentiful in the bogs and heaths. Its most outstanding botanical feature is, however, the vast and continuous carpet of woolly fringe moss which extends for mile after mile along its broad upper spurs and summit ridges. There are soil hummocks and ridges in plenty as evidence of frost action, but the moss cover is almost unbroken, and the high ground lacks the bare, stony areas of the Fannich and Beinn Dearg summits to westward. Between Wyvis and these Loch Broom hills lies a wild stretch of lower moorland, boggy in part and seldom penetrated by the naturalist. Drier, heathery moors around the Struie road have numerous patches of willow scrub, and in places, vivid patches of green grassland amongst the heather are a spectacular demonstration of how application of fertilisers, together with the sowing of grasses, can radically change the upland vegetation and enhance its value to grazing animals.

Fig. 67 Ungrazed cliff ledge covered with luxuriant growths of fern, especially alpine lady fern, Beinn Bhan, Applecross, Ross-shire.

Fig. 68 Sea campion, a common plant of coastal cliffs and shores, but found inland on some mountains.

The Islands

The Western Isles form a far-flung archipelago extending a distance of 208 miles from Islay in the south to Lewis in the north, and varying in size from insignificant rocky skerries to large islands with major mountain systems. The Atlantic Ocean gives a prevailing mildness of climate sufficient to encourage attempts at the commercial growing of tulip bulbs in the southerly isle of Tiree, and to allow the existence of a native flora containing many frost-sensitive plants.

Island floras are generally less rich in number of different kinds of plant than those of adjoining mainlands, because the smaller areas of the offshore islands reduce the chances of a plant ever arriving there by natural processes of migration, or of surviving if conditions change for the worse. The Western Isles are no exception, and whilst the sea is no barrier to plants which spread by tiny, wind-blown spores—the ferns, mosses, liverworts, lichens and fungi—the flowering plants, with their heavier fruits and seeds, are somewhat less well represented than on the mainland of the western Highlands. The flora is nevertheless quite rich, and at lower levels provides some of the most colourful displays to be seen in the whole of the region.

The visitor to the Outer Isles, arriving by plane at Benbecula, descends into a strange, watery landscape—a dark, barren, peat-clad moorland honeycombed by a myriad of lochans and lochs of all shapes and sizes, and broken by numerous protruding bosses of the underlying gneiss. On either side, the interiors of South and North Uist present much the same desolate appearance, whilst farther north the Isle of Harris is even more a sterile desert of rock (*Fig. 62*), and Lewis a wilderness of peat bog.

Over large areas of this rain-soaked land, about twenty different kinds of flowering plant make up almost the sum total of vegetation. Yet along the western coasts of this island chain, especially the first three, are broad strips of contrasting fertility and floral variety, the famous machair. Where these Western Isles present a low coast-line to the west, the violence of the prevailing westerly, on-shore winds has driven large quantities of sand on to the land. There is usually a fringe of unstable sand dunes at the head of the beach, and in a few places, as at Baleshare in North Uist, these sand hills stretch well inland, with their intervening hollows forming damp 'slacks', much as in many of the large dune systems around other parts of the British coast. More typically, though, behind the first dune ridge, the ground flattens out into a level or gently undulating plain of stabilised sand, dry for the most part, but giving way here and there to marshes or even shallow sheets of open water. This is the true machair, forming a coastal strip from a few hundred yards to a mile in width. It covers much of the low-lying island of Tiree, and the western coasts of Barra, South Uist, Benbecula and North Uist, but elsewhere occurs in only scattered localities and rather small areas.

The sand which forms the machair contains a fairly high proportion of grains derived by the grinding, under wave action, of the shells of marine animals. These shells are formed largely of lime, calcium carbonate, so that the ground-up fragments form a calcareous sand, which gives the machair its fertility and botanical variety. Wherever it occurs, the machair is part of the crofting lands, and in any one place its use has fluctuated between arable and pasture. Most machairs have at times been ploughed and sown to crops such as oats and potatoes, and then allowed to revert through the natural colonisation of common grasses and other herbs. All stages in this transition can be seen in places, but the botanically exciting machairs are those which have had longest under a regime of cattle and sheep grazing. They are essentially grasslands, but with a profusion of other herbaceous plants which in late spring and early summer give splendid changes of colour to the emerald green of the nutritious sward.

The flora of one machair is very much like that of the next, but different plants tend to be prominent from place to place, and to give the particular quality of colour. Typically there is a wash of yellow, but this can be produced by buttercups, ragwort or primroses. Some machairs are white with a profusion of daisies. Other familiars are the dandelion, yarrow, white clover, bird's foot trefoil, kidney vetch, ladies' bedstraw, wild carrot, knapweed, wall pepper, milkwort, self heal, ribwort plantain, yellow rattle, wild thyme, eyebright, wild pansy, storksbill, several mouse-ear chickweeds, and carnation sedge. The lesser meadow rue is a more unusual plant, but there are rather few rare species.

Some machairs have damper areas and even shallow lochs. The calcareous water here gives an aquatic and marsh vegetation with similarities to that of rich lowland fen. Many different kinds of sedge grow here, and there is a good deal of the handsome purple and crimson marsh orchids (*Fig. 60*). Wild iris, marsh marigold, water mint, marsh bedstraw, marsh pennywort, red rattle, ragged robin, marsh willowherb, ladies' smock, marsh cinquefoil, brooklime and horsetails. Open water has a good deal of stonewort, pondweeds of various kinds, mare's tail and bog bean.

The plant list for these machair areas is quite large, taking the range of habitats from dry to wet. Yet the influence of the blown sand does not usually extend much more than a mile inland and, where the land runs more steeply into the sea, it is lacking altogether. The barren moorlands and lochs which cover most of the Outer Hebrides are essentially similar in botanical character to those of the west Highland mainland, though some of the 'peat alpines' are less well represented. Lochs in South Uist are the habitat of the leafy pondweed, an American species confined in Britain to this area. The aquatic flora is quite limited, though, with plants such as water lobelia, shoreweed, floating scirpus, floating bur-reed and water milfoil. Some of the islands have dense growths of scrub composed of willows, hazel, rowan and juniper, with lush growths of herbs and ferns. Royal fern often grows here luxuriantly, and the tree lungwort lichen hangs in long festoons from some of the bushes. Such vegetation may once have had a much wider distribution before humans appeared on the scene, but perhaps the wetter bogs have always remained free of scrub.

Whatever the case, the Outer Hebrides are an almost treeless country now, apart from a few rocky glens fringed with scrubby woodland, such as Allt Volagair in South Uist, and planted woods, such as those around Stornoway Castle. Lewis is mainly bleak, peat-clad moorland, but Harris has some fine, rugged hills of gneiss, with Clisham (2622 ft.) the highest point. The rock is mostly acidic though, and the montane flora is rather poor, with alpine ladies' mantle, starry saxifrage, rigid sedge, least willow, roseroot, mountain sorrel, alpine meadow rue and viviparous bistort as the more noteworthy species. Some of the cliff ledges are smothered with great lush banks of greater woodrush.

The best places for alpine plants in the Hebrides are on the basaltic rocks of Skye, Rhum and Mull. The gabbro mountains of the Black Cuillin on Skye are magnificently precipitous and spiry, but their vegetation is extremely sparse at the higher levels. Some of their cliffs have a modest alpine flora, and one corrie has the distinction of being the sole British locality for the alpine rockcress, a white-flowered crucifer. The granite hills of the Red Cuillin are still less rewarding, and by far the best ground is on the great east facing escarpment of the Trotternish ridge in northern Skye. This includes Ben Edra, the Storr and the Quiraing (*Fig. 63*) with their incredible rock scenery produced by massive landslips. Amongst the strange rock pillars and towering cliffs there occur many of the lime-loving alpines mentioned for the Beinn Dearg massif, some growing in great abundance. In bare, gravelly flushes above the cliffs are other interesting plants, chief among them the inconspicuous little Koenigia, an Arctic relative of the sorrels, which was discovered here in 1934. Specimens collected on the Storr were not recognised as this new British species until 1950.

There are quite fertile hay meadows and pastures with a lowland flora along the coastal strip below the Trotternish basalt ridge, and in the southern part of Skye are interesting birch woods and hazel scrub which are rewarding to the student of mosses and liver-worts. One birchwood on the limestone band crossing Strath Suardal has mountain avens growing within its edge. There are other good plants on this southernmost outcrop of the Durness limestone, such as dark red helle-borine and holly fern, whilst the woods them-selves and an associated calcareous wetland complex are of considerable interest. The rock whitebeam has scattered occurrences on low-lying cliffs in Skye, and the island has the most northerly known localities for Tun-bridge filmy fern. It is outstandingly rich in ferns, mosses, liverworts and lichens which have a strongly Atlantic distribution in Britain. Some of the lower moorlands are not particularly interesting, but lochans near Sligachan have the rare and interesting pipe-wort. Much of the coast of Skye is cliff-bound and the flora of maritime rocks is well developed. Grassy slopes above the sea are the main habitat in Britain for the rare red broomrape, a parasitic plant associated with wild thyme.

John and Hilary Birks have made an extremely detailed study of the botany of Skye, both past and present. Their researches reveal not only the outstanding richness of the island, with 589 native species of flowering plants and ferns, 370 mosses, 181 liverworts and 154 lichens, but also the great antiquity of some of the most distinctive vegetation types, which have persisted more or less unchanged for thousands of years in places where human disturbance has been negligible.

The adjoining Isle of Rhum is a splendid place for the naturalist, for within a fairly small compass there is a tremendous range of conditions supporting a most varied flora and fauna. There are quite high mountains and the geology is very diverse. The ultra-basic rocks forming the highest hills are relatively soft and have yielded deep soils which are easily eroded by wind and water, giving rise to strange desert-like landscapes of fell-fields and rock.

There are interesting outlying localities here for rare plants such as Arctic sandwort, alpine saxifrage, two flowered rush, alpine penny-cress and forked spleenwort, and some more widespread mountain plants such as northern rockcress, purple saxifrage, moss campion, roseroot and mountain sorrel grow in some abundance. There is also a good deal of the northern bugle, which is so strangely absent from Skye. Royal fern is profuse on many sea cliffs, upright bitter vetch grows in several places and the stone bramble is remarkably abundant on some of the fell-field debris. Black bog-rush is in tremendous quantity in the bogs and more especially the flushes, but bog myrtle is exceedingly scarce, for no apparent reason except chance.

But something of a cloud hangs over the botany of Rhum. Some 30 odd years ago, a strange list of plants was reported to have been found on the island. A few were new to Britain altogether and the rest were rarities, ranging from plants known only on the central Highlands mountains to others confined to south-west England. Some of these plants undoubtedly grew on the island, but virtually all of them seem to have disappeared. The consensus of informed botanical opinion is that they were deliberately planted. Most of the dubious records have been dropped from the literature, and it is fairly clear which of the rarer plants are truly native to Rhum. The episode underlines the great confusion which can be created in the study of the natural processes of plant distribution by the deliber-ate introduction of species to new places, however well-intentioned such actions might be.

Still farther south, the large island of Mull

has a rich flora. There are more basalt mountains with notable plants, such as the little Koenigia in its second British locality, again in bare, moist spreads of gravel; mountain avens, rock whitebeam and northern rock-cress. Coastal rocks have an isolated northern locality for blue fleabane, and rocky grassland has several colonies of the red broomrape. Some of the south facing escarpments have large areas of hazel scrub beneath the cliffs, and there are interesting woodland plants such as wood vetch, giant bellflower, hemp agrimony and pendulous sedge. Mull is within the range of native oak, and there are some coastal woods with an abundance of this tree, though birch is more prominent in most broad-leaved woodlands. In common with the other larger islands, much of the interior of Mull is rather ordinary hill country and moorland with a flora differing little from that so characteristic of much of the uplands of the west Highlands mainland. The British Museum (Natural History) have made an extremely detailed survey of the natural history of Mull and their work is now in course of publication.

Islay and Jura are the southernmost of the larger islands, apart from Arran already described, and in some ways they form a contrasting pair. Islay has a good deal of low-lying ground, and its rocks have basic schists and limestones locally which give rise to lime-rich soils and waters, whilst the coast has areas of dune and ground influenced by blown sand. Jura is a narrower island with much quartzite, forming sterile and locally rugged moorland, and the famous higher peaks known as the Paps of Jura. Both islands have a good deal of fringing coastal woodland of oak, birch and hazel, though this is mostly rather poor and narrow. The climate is mild and oceanic as reflected in the great abundance of the hay-scented buckler fern in rocky woods and on shady coastal banks, and in the presence of that characteristic feature of western Ireland, the planted fuchsia hedge. There are interesting low-level marshes, especially on Islay, where rich fen has recently been found with the marsh fern, otherwise unknown in the Highlands and Islands. Other southern plants of fen and marsh include the fine saw sedge, panicled sedge, marsh St. John's wort, lesser skullcap, bog pimpernel and in one place, the meadow thistle.

The occurrence of this last plant is most interesting, as it is widespread through Ireland, but in Great Britain has no other locality north of Yorkshire. Somehow it has managed to jump the 25 mile gap between Ireland and Islay, but whether the downy seed blew there, was carried by birds, or arrived in some other way will probably never be known.

The mountain flora of the two islands is extremely limited, with only a few plants which can grow on acidic rocks and soils, such as alpine ladies' mantle, dwarf juniper and starry saxifrage. A distinctive coastal feature in places is the raised beaches, sometimes of smooth rounded pebbles without a trace of vegetation other than stone-encrusting lichens. There are associated cliffs, originally cut by the sea but now standing well above high tide mark, and grown with scrub in places. Their damp, shady caves are often richly draped with curtains of fern, moss and liverwort.

The long Hebridean chain includes many smaller islands—Colonsay, Tiree, Coll, Eigg, Muck, Canna, Barra and Raasay, to name but a few. None of them has any plant which cannot be seen elsewhere in the area, but their aggregate interest is high, and a rewarding holiday could be spent on any one of them, searching carefully to see what the total score of wild plants might be. The prospects for new finds are perhaps less than on the bigger islands or the mainland, but who knows? Coll has the pipewort in its lochs, and several of the smaller islands have the northern bugle. Coastal habitats are naturally well developed on these lesser isles, and some are wholly and severely subject to the influence of the sea. Sandy and shingly beaches with a typical strand flora abound, passing here and there at the mouth of streams into little patches of salt marsh. Some places have dunes and machair, or there are rocky shores and cliffs. Just above high tide level are the characteristic brackish marshy pastures (*Fig. 59*), densely grown with iris and rushes, and variously merging farther from the sea into drier grassland, heath or bog, or into patches

Fig. 69 White water lilies, common in the open water of northern lochs.

of scrub in protected places. Some of the small, storm-swept islands are covered largely by grassland, for the salt spray has an enriching effect, and sheep or cattle do well here, or there may be large seabird colonies whose guano has a powerful fertilising effect on the soil. The island-going botanist will nearly always find a wealth of interest without going far from the shore, and even if the rare or unusual does not turn up, there is a colourful charm in the commonplace. Daisies and dandelions are beautiful flowers if one can look at them anew, as on a remote shore lapped by the great Atlantic waters.

Seaweed communities are well represented all around the rocky coasts of the Highlands and Islands, and in some places, such as the Outer Hebrides, they form the basis for a local occupation. On the boulder beaches and rock platforms of the zone between tide levels, and extending down into the shallows below low water mark are luxuriant beds of brown algae. The various kinds of wrack, bladder, knotted and serrated grow in great quantity, attached to the rocks (*Fig. 71*), and the channelled and flat wracks are locally abundant. The largest seaweeds are, however, the kelps or laminarians which form 'forests' in the zone immediately below low water mark. They have thick rubbery stems producing a cluster of long, blade-like fronds at the apex. Heavy wave action, especially during storms, tears away masses of these seaweeds, which are washed up on the beaches. These strand-line accumulations and growths exposed at low water have for long been gathered by local people for use as fertiliser on their ploughed croftlands, which in the Hebrides includes the machair. This use still continues here, but there are also processing factories in the Outer Isles and near Oban where the dried 'tangle' (mainly stalks of kelp) is used to extract alginic acid which has a wide use in the food, pharmaceutical and textile industries. The seaweeds were formerly burned to extract iodine, soda and potash, for they are rich in nutrients and both sheep and deer often make their way to the shore to graze them. Smaller seaweeds abound on these rocky shores, with edible kinds such as dulce and Carrigheen, and beautiful feathery red algae. Above the zone of seaweeds is one dominated by lichens adapted to saline conditions. There are colourful orange patches of *Xanthoria parietina*, grey and black encrusting species, and pale green festoons of *Ramalina scopulorum*.

These rocky coasts with their algal communities and associated fauna are one of the richest and most fascinating wildlife habitats in the whole Highlands and Islands. They are the most natural of all the major habitats, yet, sadly and ironically, are now perhaps the most threatened of all, at least on the north and east coast, through the ever-present danger of a major North Sea oil spill.

Orkney and Shetland are the last of the main island groups, geographically separate from the Hebrides, but somewhat similar biologically. These two northernmost groups are rich in interest for the naturalist but their birdlife is more remarkable than their plants. They have been botanised rather more thoroughly than much of the mainland of Scotland, and we owe much to the devoted efforts of Elaine Bullard in Orkney and Walter Scott in Shetland.

Both Orkney and Shetland show 'island effects' in having a more restricted flora than that of the adjoining mainland, and many plants otherwise widespread in the Highlands are unknown on one or both of these groups. Neither has high mountains, and the great total length of coastline, much of it cliff-girt, is their outstanding physical feature.

Their climate is of the extremely oceanic kind, cool and windy, and expressed in the virtual absence of native woodland. It is not simply that human activity has eradicated forest: pollen analytical studies show that tree cover has always been sparse and patchy here, right through the post-glacial period. Little tangles of scrub with birch, rowan, willow and aspen, honeysuckle and wild roses survive in sheltered rocky places by the sea, on islands in lochs, and in narrow, rocky glens. Among the planted trees, sycamore seems best suited to the adverse climate. Yet the winter temperatures are still high enough for the warmth-loving hay-scented fern to grow in Orkney.

Orkney is formed largely of Old Red Sandstone which in places contains a good deal of lime and breaks down to give quite fertile soils. In consequence, much of the lower-lying ground in this archipelago is farmed, and is a green, productive land. By contrast, Shetland is much more complex geologically, with a preponderance of rocks which give acidic soils, except on Unst and Fetlar where serpentine outcrops extensively. Even so, the two island groups share many plant communities in common, and their floras show a good deal of similarity.

The maritime habitats and vegetation in particular are very much alike. Patches of salt marsh have typical plants such as glasswort, creeping sea grass, thrift, Gerard's rush, sea plantain, sea milkwort, sea arrow grass, sea spurrey, scurvy grass and saltmarsh flat-sedge.

Shingle and sandy beaches have open growths of Babington's orache, sea rocket, sea purslane and field milk thistle. The oyster plant belongs especially to this habitat and is widespread in these islands. One beach in South Ronaldsay, Orkney, has such a massive colony of this lovely plant that it shows as a conspicuous blue-grey patch from the air. More bouldery beaches with rocky platforms have an abundance of scentless mayweed, curled dock, sea campion, thrift and lovage. Plants such as silverweed, nettle, chickweed and common forget-me-not are often abundant where the beach tails off into less distinctly maritime ground.

Narrow dune ridges occur locally and have great quantities of the handsome lyme grass, whilst sea couch grass, marram and sand sedge are all abundant plants of the open sand. On the relatively stable sandhills there are species such as bird's foot trefoil, white clover, red fescue, heath violet, ragwort, and sometimes carpets of moss. Machair is quite well developed in places, and has many of the plants mentioned for this habitat in the Outer Hebrides. Especially attractive flowers are those of the purple and crimson marsh orchids, ladies' bedstraw, felwort, wild thyme and grass of Parnassus, and less conspicuous but interesting plants are the frog orchid, adder's tongue fern and lesser clubmoss.

The seacliff vegetation is very much the same as that of the mainland and the Hebrides. Above the tidal zone of seaweed beds is a spray-drenched belt grown mainly with lichens, ranging in form and colour from the bright orange patches of Xanthoria to the long stringy festoons of Ramalina. The steep, exposed faces above have an open growth of red fescue, sea plantain, buck's horn plantain, thrift, sea campion and lovage (*Fig. 68*), with the sea spleenwort often common in crevices and caves. On the bigger ledges and more sloping faces higher up or farther back from the sea, there is a change to the flora of ungrazed woodland, though here usually without the trees. Greater woodrush, red campion, angelica, hogweed, primrose, sorrel, foxglove, lady fern and broad buckler fern all flourish in such places. The roseroot is profuse on some cliffs, but two plants common in this habitat elsewhere do not occur in Orkney or Shetland, the bluebell and English stonecrop. The heavily manured seabird cliffs are, as usual, clad with profuse growths of scurvy grass, sorrel and grasses. Spray-drenched cliff summits have dense swards of plantain, thrift and grass, rather like salt marsh high above the sea. In places on the cliff tops are grass-heaths with great abundance of vernal squill, and in Orkney they often have the beautiful Scottish primrose as well.

The farmed areas have a goodly assortment of the common 'weeds' of the lowland agricultural environment of the Highlands, with roadside verges, permanent grasslands and arable field edges. The discerning botanist will soon realise that some characteristic species are absent or scarce, but some others compensate in their super-abundance. Primroses are especially plentiful along roadsides in Orkney, for example. There is also a good deal of marginal ground, where this flora mixes with that of the open moor. Much of the lower moorland edge has been cut for peat and the ground generally disturbed in both archipelagos.

Marshes and fens occur quite widely on the lower ground, and noticeably abundant plants here are iris, meadowsweet, ragged robin, cuckoo flower, marsh marigold, marsh pennywort, lesser spearwort, water mint, marsh cinquefoil, bogbean, red rattle and various kinds of rush, sedge and horsetail. Lochs of small to medium size are numerous and range from those with base-rich or even brackish water, similar to the Hebridean machair lochs, to the typical inky dubh lochans of the sterile blanket bogs.

There are good areas of blanket bog on the higher moorlands of Orkney, but this kind of ground is especially extensive in Shetland. Most of the island of Yell is covered by it—a dark, barren mantle of peat, desolate as the tundras of the far north. The plant life of such places is as limited as usual, but two plants are noticeably profuse in these northernmost British blanket bogs—the common cotton-grass, with its branched flower-heads, and the woolly fringe moss, forming dense cushions and carpets.

In places this ground is colourful with hummocks and patches of Sphagnum, green, yellow, orange and red, but the patterned pool systems of the mainland bogs are hardly developed here. Bog extends to the cliff tops, or almost to sea level in places, but the drier uncultivated ground has a good deal of heath, dominated essentially by short ling heather, but often with a considerable mixture of other plants. Crowberry, bell heather and creeping willow are other common shrubs, and near the sea there is an abundance of herbs such as tormentil, bird's foot trefoil, devil's bit scabious, heath spotted orchid, heath bedstraw, lousewort, plantains, self-heal, thyme, heath violet, eyebright, beautiful St. John's wort, daisy, autumnal hawkbit, mountain everlasting, grasses and sedges. In places this maritime heath is replaced by communities with mat grass, heath rush or sedges. Heather is mainly of a rather short growth in these northern isles, but there are rank growths on some of the hills of Hoy, which have been little burned in recent times. Bilberry heath is rather sparse everywhere in these islands.

The altitudinal descent of vegetation zones noted in the north-west of Sutherland, by comparison with the central Highlands, is even more marked in Orkney and Shetland. On some of the west coast headlands here, at elevations as low as 400 ft., the typical maritime heaths just described are replaced by the montane kind, rich in dwarf shrubs such as dwarf juniper, common bearberry, alpine bearberry, dwarf azalea, bog bilberry and least willow. Fine examples occur on the moor near the Old Man of Hoy in Orkney (*Fig. 64*).

Perhaps even more remarkable is the incredible amount of lichens of the 'reindeer moss' group in the heath community near the summit of St. John's head. Lichen heaths of this type are found mainly at fairly high levels on the mountains of the eastern Highlands where the climate is somewhat continental, though examples also occur on coastal moors of north-west Sutherland.

Both Ward Hill on Hoy and Ronas Hill in mainland Shetland have finely developed systems of downslope terraces and wind-stripes amongst their alpine heaths. Such features are normally found at much higher levels, where low temperatures and high winds are frequent. The summit of Ronas Hill, not quite reaching 1500 ft., is one of the most spectacular fell-fields in the whole of Scotland. There is a sea of granite blocks, with spreads of gravel and sand, even more sparsely vegetated than the tops of the high Cairngorms. Amidst this stone desert are curious little rosettes of alpine saw-wort and sea plantain, along with mountain everlasting.

The higher hills of Hoy have some inland calcareous sandstone crags with a good alpine flora which includes mountain avens, holly fern, purple and yellow saxifrages, hoary whitlow grass, alpine saw-wort, alpine meadow rue, viviparous bistort, moss campion, rose-root, mountain sorrel, whortle willow and alpine meadow grass. This is the most northerly British locality for some of these plants. Good calcareous rocks are fewer in Shetland, but the far north of Unst has something special to offer. Serpentine, a rock with high magnesium content, occurs extensively here, but is mostly covered with a species-rich heath similiar to that described.

Near Baltasound, however, is another fell-field, at a much lower level even than Ronas

Hill, and consisting of open serpentine debris. Here grows Edmonston's Arctic mouse-ear, in its only British station, along with Norwegian sandwort, northern rock-cress, moss campion and yellow saxifrage. Edmonston was a brilliant young Shetland botanist whose career was tragically cut short by an accident at sea. His plant survives, though reclamation of part of its habitat by the addition of fertiliser and grass seeds mixture has been surprisingly successful, and shows how even apparently quite barren ground can be made to produce, given the right treatment. The secret of the survival of the rare northern plants here may indeed lie in the natural poverty of the rock in certain nutrients, which excludes normally vigorous, competitive plants and maintains the open conditions needed by the others. There is another long-lost alpine hereabouts, in the mountain sandwort.

Beyond Baltasound, peaty headlands stretch away to the north coast of Unst, northernmost point of the British Isles. This is an appropriate point to end our brief scan of the plant life of the Highlands and Islands. It is a country through which the naturalist could roam for the whole of a lifetime without fear of exhausting its interest and to which the visitor will be drawn back again and again. We must hope that it will forever remain a region of magnetic beauty and character, rich in wild plants and animals, but this cannot be taken for granted. The last short section of this book is accordingly directed towards the prospects and the problems.

Fig. 70 Least willow, one of the abundant mountain plants of poor soils.
Extreme right
Fig. 71 Bladder wrack, a common seaweed of rocky shores.

The Future

This book is published by an official body set up to promote development in the Highlands and Islands. Britain is an overcrowded country in which economic survival has become a perennial worry, and it is inevitable that people look keenly at any large area of seemingly undeveloped land to see what advantage may yet be wrung from it.

The Highlands and Islands are, in fact, already an exploited land—some would say over-exploited—and the bogey of rural depopulation could be in part an expression of a declining carrying capacity, though social factors are perhaps more important still. The great forestlands of the past were exploited in a largely extractive way. There followed a period of sheep and game cropping sustained at least in part by imported capital, but extractive in the sense that virtually no attempt was made to balance the removal of nutrients (in animal bodies) from the land. Sheep farming is a tenuous business nowadays even when heavily subsidised, and a great deal of the poorer sheep ground has been sold for forestry enterprises during the last few decades. This has gone some way to restoring the lost woodland cover though the new forests are mostly of totally different character from the old. Some of the best-run estates managed to have a well-balanced combination of sheep, forestry and game management. The game element, of deer, grouse and salmon, depends largely on private wealth and, despite the influx of overseas interest, is under increasingly adverse pressure. The potential for arable farming is strictly limited, by climate; and though techniques of reclaiming marginal land for pasture are well established, their wider application is limited by cost. Similarly, attempts to increase the profitability of hill cattle, to exploit the red deer better as a range or even domesticated animal, and to eradicate the large areas of bracken, all come up against the hard economic facts of life and background constraints of EEC agricultural policy.

These are the resources of soil and living organisms. After them are the inorganic raw materials. The Highlands have an excess of water, most of it is too far from centres of need to be usable as water supply, but the North of Scotland Hydro-Electric Board, through a vast programme of engineering works, has tapped a substantial fraction of the mainland catchments to generate hydro-electric power. This power has been used in the smelting of aluminium at Kinlochleven, Fort William and Invergordon.

Mineral working in the past has been on a small scale, but there have during recent years been systematic and intensive surveys of mineral deposits in many parts of the Highlands and Islands. Geologists seem—perhaps with good reason—not to speak with one voice on the prospects for actual exploitation, but the portents are rather obvious.

One major resource is only too well known, and its development is having considerable impact on some parts of the area—North Sea oil and gas rig construction yards have been built in several places, and there is a growing paraphernalia of terminals, storage tanks, pipelines and so on. The wrangle over the oil refinery at Nigg Point merely postponed the inevitable. Industry has come to the Highlands, and it is likely to beget more industry.

Finally, there is recreation and tourism, which some see as the greatest untapped potential of the Highlands and Islands. The Aviemore development has shown what can be done, and there is much enthusiasm for expanding facilities and enhancing the attractions of holidays as a means of providing employment and drawing in wealth.

What relevance has all this to the plant life of the Highlands? It has many implications for change. When development takes place, wildlife is usually the loser. Afforestation in the modern style is a good example, for under conifers the other vegetation is often virtually eliminated. The loss is especially serious when native, broad-leaved trees are replaced by exotic species. On average, grouse moors have a more interesting vegetation and associated fauna than sheep walks, but the pressure to utilise the uplands to the full as grazing land may result in a gradual loss of interest in grouse, and reclamation of unenclosed hill ground of various kinds will continue. There is likely to be a heavy encroachment on existing marginal land by way of agricultural improvement. Hydro-electric and industrial developments tend to have an obliterating or severely damaging effect on botanical habitat. Mining and quarrying can have similar effects, but the abandoned workings sometimes develop a different kind of interest. The need for access routes and construction material, road improvement schemes, and various other facilities all have their impact. Not least in influence are the tourist and recreational developments, which have already resulted in the invasion of certain remote hill country, notably in the Cairngorms, which was once only accessible to the determined and the fit. Ski developments and chairlifts have already left their unsightly scars, and the problems of erosion associated with too many trampling feet are beginning to develop in the manner all too apparent in the much trodden mountains of Lakeland and Snowdonia.

A sensitive yet firm planning control will be needed to cope adequately with the impending problems. It is far too early to assess whether present arrangements are adequate, but the existing virtual exemption of agriculture and forestry from planning responsibility must necessarily limit what might be achieved in this direction.

The most important of the areas of special botanical interest are already known to their owners and the planning authorities. I would hope that these can receive adequate protection and management, where this does not exist already. The future of the remaining and much larger part of the Highlands and Islands lying outside such areas is much more problematical. The conservation of its wild flora and fauna will depend very much on the attitudes of the owners, occupiers and other residents, the prospective developers, and the holiday visitors. There are bound to be further losses amongst wildlife, but their extent and severity will depend on how much people care and behave.

My hope is that this book may help towards a greater appreciation of the beauty and interest of the wild plants of the Highlands and Islands, and a sense of need to cherish them. My worry is that it may cause greater disturbance to some of the places where they grow and, more particularly, to still further collecting of the rarer or showier kinds.

If this last is the case it will have done more harm than good. So I close with a plea for respect for this most fragile of the natural resources of the Highlands and Islands. Let us think of it as Nature's laboratory, museum and art gallery—all rolled into one, where people can come to study, learn and admire for as long as humanity has an interest in such things.

General Index

Selected Plants Index